OPPORTUNISTIC ADAPTATION
USING THE URBAN RENEWAL CYCLE TO ADAPT TO CLIMATE CHANGE

OPPORTUNISTIC ADAPTATION
USING THE URBAN RENEWAL CYCLE TO ADAPT TO CLIMATE CHANGE

DISSERTATION

Submitted in fulfilment of the requirements of
the Board for Doctorates of Delft University of Technology
and
of the Academic Board of the IHE-Delft
Institute for Water Education
for
the Degree of DOCTOR
to be defended in public on
Wednesday 28 April, 2021 at 10.00 hours
in Delft, the Netherlands

by

Polpat NILUBON

Master of Excellence in Architecture,
The Berlage Institute, Netherlands

born in Bangkok, Thailand

This dissertation has been approved by the
promotor: Prof. dr. ir. C. Zevenbergen and
copromotor: Dr. ir. W. Veerbeek

Composition of the doctoral committee:

Rector Magnificus, TU Delft   Chairman
Rector IHE Delft   Vice-Chairman
Prof. dr. ir. C. Zevenbergen   IHE Delft / TU Delft, promotor
Dr. ir. W. Veerbeek   IHE Delft, copromotor

Independent members:

Prof. dr. ir. V.J. Meijer   TU Delft
Dr. K. Leeruttanawisut   UN-HABITAT, Thailand
Prof. dr. V. Nadin   TU Delft
Prof. dr. N. Ongsavangchai   Chiang Mai University, Thailand
Prof. dr. ir. J.A. Roelvink   TU Delft / IHE Delft, reserve member

Cover illustration by Nil-u-bon (2015)

This research was conducted under the auspices of the Graduate School for Socio-Economic and Natural Sciences of the Environment (SENSE).

CRC Press/Balkema is an imprint of the Taylor & Francis Group, an informa business

Published by:
CRC Press/Balkema
Pub.NL@taylorandfrancis.com
www.crcpress.com – www.taylorandfrancis.com
ISBN 978-1-032-05509-1

This dissertation is dedicated to my guardian angels in heaven

Uraiwat, my grandmother

And

Likit sir, my Architect guru

Summary

Urban climate adaptation is increasingly being conceived as a bottom-up socio-economic process, taking a dynamic view on adaptation by combining climate change with socio-economic drivers, rather than as a set of stand-alone adjustments. This approach of adaption has been referred to as 'mainstream adaptation' and often results in an additional intervention/component which is integrated or embedded in (re)development projects. Within 'mainstream adaptation', a new term referred to as Opportunistic Adaptation (OA) has been introduced in this dissertation. The term refers to seizing climate change adaptation opportunities exclusively during the renewal or regeneration process of city components. This autonomous process provides a window of opportunity to adapt/adjust the existing urban fabric. While the need for adaptation mainstreaming has been identified by many authors, operational methods at the neighborhood level are currently lacking the tools and perspectives to include urban dynamics. In particular, this holds true for those tools which support the design process from an architectural or urbanism point of view.

This dissertation aims to provide insight into the potentials of mainstream adaptation, with particular attention to the renewal cycles of the components of the urban fabric (such as building construction and infrastructure). These insights are required to assess the Adaptation Rate (AR) and to explore and identify adaptive measures to be implemented across different spatial scale levels (from building to neighborhood level). In this research a range of methods has been developed, applied and combined, such as a method to develop a decision tree to sequence adaptation interventions (adaptive pathways), a method to spatially map opportunities to intervene in the urban fabric and a method to assess the Adaptation Potential (AP).

As opposed to the traditional top-down and standalone adaptation interventions, OA focusses on urban services in an integrated way by considering various spatial scales (ranging from individual building to landscape level), which allow to identify multiple benefits and lead consequently to an improvement of the overall urban performance. OA requires strategic planning and long-term thinking, where the purpose of a future projection (based upon the lifecycle and the renewal rate of individual urban components (referred to as assets)) is to develop adaptation pathways consisting of a sequence or sequences of consecutive adaptation interventions of buildings and infrastructure components located in a flood-prone area. The identification and selection of adaptation interventions are important steps in the process as they impact the future form and space of the urban landscape as well as the performance of the area as a whole. Moreover, they also determine to a large extent the flexibility of the adaptation strategy: some interventions may lead to lock-ins, while other leave room for adjustments in the future.

This dissertation describes a 'action research' approach which aims to develop and apply OA in a real-life context. The underlying main objectives of this dissertation are (i) to define and assess the AP based on renewal cycles of urban components; (ii) to identify adaptive pathways based on a sequence (or sequences) of interventions emerging from mainstream adaptation of buildings and infrastructure; and (iii) to develop the contours of an adaptation strategy which take into account co-benefits as well as future flexibility associated to specific adaptation measures along the respective pathways for an urban area of the city of Bangkok.

The operational 'philosophy' to engage stakeholders in developing the OA approach which has been followed in this research, can be best described as participatory action research. Its purpose is not (primarily) to advance scientific knowledge but to achieve and improve practical applications that address real problems. The action research methodology is structured around three parts. Firstly, a framework has been developed to integrate the process of urban renewal and adaptation at the scale of a single building to a neighborhood. Secondly, design principles have been identified and supporting tools developed to produce dynamic designs based on adaptation pathways. Thirdly, workshops have been organized to actively engage stakeholders and end-users in the refinement and demonstration of the resulting OA approach using a case study area and to conduct an ex-ante evaluation.

This research reveals that a long-term, strategic approach to climate adaptation of cities is needed. This strategy should be based on an understanding of the urban dynamics and on the flood risk of an urban area. An OA approach, such as developed and demonstrated in this research, is required to take advantage of the AP of an urban area and to maximize the associated multiple benefits of adaptation interventions. Adaptation pathways seem to be well suited for this purpose given the support expressed by the stakeholders during interactive sessions demonstrating the OA approach in a real-life context. Further research will be needed to assess the costs and benefits of the proposed adaptation interventions and their effectiveness, which are beyond the scope of the dissertation. This will require additional, detailed flood risk modelling of the case study area for the current situation (present time) and future projections (using various scenarios).

Samenvatting

Stedelijke klimaatadaptatie wordt steeds vaker gezien als een "bottom-up"-gedreven sociaaleconomisch proces, waarbij sociaaleconomische en klimaatgerelateerde ontwikkelingen worden gecombineerd in plaats van afzonderlijk benaderd. Deze benadering van adaptatie wordt vaak getypeerd als "meekoppelen" en resulteert vaak in het plegen van additionele interventies die worden geïntegreerd in ontwikkelings- en herontwikkelingstrajecten. Het concept van meekoppelen wordt in deze dissertatie uitgebreid met de introductie opportunistische adaptatie (OA). Deze term refereert naar het benutten van mogelijkheden voor klimaatadaptie enkel op het moment van ontwikkeling of herontwikkeling van stedelijke componenten (bijv. gebouwen, wegen, etc.). Juist de cyclus van stedelijke autonome proces biedt kansen voor het aanpassen van het stedelijk weefsel aan klimaatverandering. Hoewel er overeenstemming is over de noodzaak tot meekoppelen zijn er maar weinig operationele methoden die voldoende handvaten en methodes bieden om een dergelijke benadering te integreren in stedelijke ontwikkelingsdynamiek. Dit geldt met name voor methodes die het architectonisch of stedenbouwkundig ontwerpproces ondersteunen. Deze dissertatie is gericht op het identificeren van de potentie ván het meekoppelen van klimaatadaptatie, waarbij vooral nadruk wordt gelegd op de cycli van herontwikkeling van componenten in het stedelijk weefsel. Inzicht hierin is een voorwaarde voor inzicht in het adaptatietempo en het identificeren en erkennen van adaptatiemaatregelen voor verschillende schaalniveaus (van gebouw tot buurt). In dit onderzoek is een scala aan methodes ontwikkeld, toegepast en gecombineerd, zoals een methode om een beslissingsboom te ontwikkelen met combinaties an adapatiemaatregelen (adaptatiepaden), een methode om een kaart te vervaardigen waar de kansen voor interventies (kansenkaart) zijn aangegeven en een methode om het adaptievermogen te bepalen.

In tegenstelling tot traditionele top-down adaptatie geformuleerd in afzonderlijke projecten, focussend OA op het bieden van meerwaarde geïntegreerd in het betreffende schaalniveau. Dit zorgt uiteindelijk voor beter functioneren. OA vereist strategisch plannen vanuit een langetermijnperspectief waarbij de nadruk ligt op het creëren van projecties bestaat uit het definiëren van adaptatiepaden gebaseerd op de levensduur en het herontwikkelingentempo van stedelijke componenten. Deze bestaan uit een sequentie van maatregelen op het niveau van gebouwen en infrastructuur in overstromingsgevoelige locaties. Identificatie en selectie van maatregelen zijn een belangrijke stap in dit proces daar ze bepalend zijn voor de vorm en ruimtelijke ontwikkeling van het stedelijk weefsel alsmede hun functioneren als overstromingsbestendige gebieden. Daarnaast, bepalen de adaptatiepaden in hoge mate de flexibiliteit van de adaptatiestrategie: sommige maatregelen zijn beperkend terwijl andere juist ruimte creëren voor het kiezen van maatregelen in de toekomst.

Deze dissertatie beschrijft een "action research"-benadering welke als doel heeft om het concept van OA te ontwikkelen en toe te passen in een werkelijke situatie (casus). De onderliggende hoofddoelen van deze dissertatie zijn: (i) het definiëren en bepalen van het adaptatievermogen gebaseerd op de levensduur en levenscycli van de stedelijke componenten, (ii) het identificeren van de bijbehorende adaptatiepaden waarmee de volgorde van interventies in beeld zijn gebracht voor gebouwen en infrastructuur en (iii) het in beeld brengen van de contouren van een adaptatiestrategie die de meervoudige voordelen van adaptatie in beschouwing neemt alsmede de flexibiliteit van adaptatiemaatregelen als onderdeel van adaptatiepaden voor een stedelijk gebied in Bangkok. De operationele benadering om stakeholders te betrekken bij de ontwikkeling van de OA-benadering die is gevolgd in dit onderzoek kan het best beschreven worden met participatief action research. De doelstelling hiervan is niet alleen om wetenschappelijk onderzoek verder te brengen, maar juist om de praktische toepasbaarheid daarvan te vergroten, die nodig is om de echte problemen aan te pakken. De "action research"-benadering is gebaseerd op drie onderdelen. Het eerste onderdeel betreft een raamwerk dat is ontwikkeld om het proces van stedelijke vernieuwing en adaptatie te integreren op een schaal van een enkelvoudige gebouw tot een buurt of wijk. Het tweede onderdeel gaat over de ontwerpprincipes die geïdentificeerd zijn met de daarbij behorende ondersteunende instrumenten om een dynamisch ontwerp te vervaardigen gebaseerd op adaptatiepaden. Het derde onderdeel betreft de workshops die zijn georganiseerd om de stakeholders actief te betrekken bij de verfijning en demonstratie van de OA benadering voor een casus die is gebruikt voor een ex ante evaluatie. Dit onderzoek toont aan dat een lange-termijn strategische benadering voor klimaatadaptatie van steden noodzakelijk is. Deze strategie dient gebaseerd te zijn op kennis van de stedelijke dynamiek en van overstromingsrisico's van het stedelijk gebied. Een opportunistische adaptatiebenadering, zoals ontwikkeld en gedemonstreerd in dit onderzoek, is nodig om maximaal profijt te hebben van het adaptatievermogen, inclusief de meervoudige voordelen die deze adaptiemaatregelen bieden. Adaptatiepaden lijken geschikt voor dit doel gegeven de positieve reacties van de stakeholders op deze methode tijdens de interactieve sessies, waar de OA benadering is gedemonstreerd in een casus. Verder onderzoek is nodig om de kosten en baten te analyseren van de voorgestelde adaptatiemaatregelen en hun effectiviteit, hetgeen nu buiten de reikwijdte valt van dit onderzoek. Dit vraagt om aanvullende, gedetailleerde hydrologische modellering van het studiegebied, zowel voor de huidige als toekomstige situatie (voor verschillende scenarios).

Acknowledgements

The time has come to be reminded that my long PhD journey has been carried out throughout six years at IHE Delft, the Netherlands. Without a doubt, this research would not have been possible without help and support from the kind people around me, whom of which I would like to give particular mention hereafter.

First, I thank the sponsor: Rajamagala University of Technology Thunyaburi (RMUTT), for financial support. I am very grateful to Mr. Likhit Sirichote for giving me the opportunity to grow at RMUTT.

My sincere gratitude goes to Prof. Bas Jonkman, as he was the first to introduce me to Prof. Chris Zevenbergen, which provided me with such an open Ph.D. experience at one of the best institutes for water education (IHE Delft). My sincere gratitude also goes to my supervisors, Prof. Chris Zevenbergen, for his scientific guidance, support, and motivation throughout the 6-year period of research. Chris, your views on resilience have been a key source of inspiration for this research. Thank you for advising and sharing your expertise on water-related technologies, and policies with me. Thank you also for providing critical feedback on this dissertation especially on the design section as well as for your continuous support. Finalizing this research even during New Year holidays manifested your commitment as a supervisor to my research. A grateful thanks also goes to Dr. William Veerbeek, my daily supervisor, for scientific research and his patience with my research, and for being very inspirational.

This research would not have been possible without the support and cooperation of the participants from the local community in Lad Krabang, Bangkok Municipal Administration (BMA), and the Lad Krabang district office as well as the Institute of Metropolitan Development (IMD), Thailand. For the case study fieldwork of Lad Krabang, Bangkok, I would like to thank the Lad Krabang District office,

Bangkok for providing valuable data. The kind support from Ms. Thipawan Saenchan for their advice and assistance with data collection and data processing is also greatly appreciated. Special thanks are due to Dr. Kittima Leeruttanawisut (UN-HABITAT) for making this PhD research a more enjoyable experience. I would like to acknowledge the help of Dr. Thongchai Roachanakanan (DTP) in sharing the flood situation and urban flood system information of Bangkok. I am also grateful to Dr. Supachai Tantikom (Chief Resilience Officer) and Assoc. Prof. Banasopit Mekvichai (BACC) for providing me with the urban flood strategic planning of Bangkok.

Many thanks go to all my friends in the Netherlands that made me feel at home. I would like to extend my gratitude to all of them and their families. Many thanks in particular to my direct colleagues at the former WSE Department. Of course, this PhD experience would not been the same without the stimulating company and friendship of my fellow at FRG: Carlos Salinas Rodríguez, Hieu Ngo, Maria Luisa Salingay, Minh Ha Do, Muhammad Dikman Maheng, and Nguyen Huynh as well as Dr. Mohanasundar Radhakrishnan. Many thanks to Victor Hugo Da Motta Paca (my brothers) for helping me with recommendations and all other small things. Thanks to Ms. Tonneke Morgenstond, for providing me a workplace with an unforgettable view of the Oude Delft. Many thanks to Ms. Jolanda Boots, and Ms. Anique Karsten for all their kind assistance and useful recommendations for making life in the Netherlands feel like home.

Last but not least, I want to thank to my family and friends for providing welcome distraction from my PhD research. Calin, Taesan, Hiroki, and P'Gap, thank you for believing in me, and for your love and support. And Neewalak, thank you for always being there, to support me in difficult times, and to share happiness in good times. You are the best part of my life.

Table of content

Figures and tables

List of tables

Acronyms and definition of terms

Acronyms

AC: Adaptation capacity

AP: Adaptation Potential

AR: Adaptation Rate

ATP: Adaptation Tipping Point

CBD: Central Business District

cc: Climate change

e.g.: exempli gratia, meaning "for example"

EOLC: End of the lifecycle

EOLC-Gap: End of the lifecycle gap

EOLs: End of the lifespan

et al.: et alii, meaning "and others"

FAM: Flood Adaptation Measures

FIA: Flexibility In Adaptation

FRM: Flood Risk Management

i.e.: id est, meaning "that is"

LCA: Lifecycle Assessment

RIO: Real In Option

Definition of terms

Adaptation: The process that entails responding to uncertain changes in drivers, pressures and impacts on a system.

Adaptation capacity: The capacity of a system to adjust to climate change (including climate variability and extremes) to moderate potential damages, to take advantage of opportunities, and/or to cope with consequences.

Adaptation Pathway: A sequence of responses and potential adaptations, which may be triggered before an Adaptation Tipping Point (see definition below) occurs.

Adaptation Potential: The number of opportunities to adapt buildings and infrastructure in the future arising from renewal. The lifespan of individual assets defines the potential for an adjustment over a longer time period.

Adaptation Tipping Point (ATP): A physical boundary condition where acceptable technical, environmental, societal or economic standards may be compromised.

Approach: the main orientation of the climate impact and adaptation assessment.

Flexibility: The ease or difficulty with which a system or a component of the system can be adjusted (e.g., time and options) to future change.

Flood Risk Management system: The whole of the physical systems, actors and rules required to manage flood risk.

Maladaptation: An action taken supposedly to avoid or reduce vulnerability to climate change that impacts adversely on, or increases the vulnerability of other systems, sectors or social groups.

Measure: A measure that aims to reduce risks by modifying the system through physical and built interventions.

Method: A systematic (i.e. stepwise) process of analysis.

Opportunistic Adaptation: A climate adaptation approach that is based on an understanding of the AP and in which adaptation measures are integrated into urban renewal cycles.

Real In Option: A Real Option created by changing the engineering system (re)design.

Resilience approach: A dynamic perspective on adaptive processes and the effects of these processes at/across different spatio-temporal scales.

Scenario: Plausible and internally consistent view of the future, which is used to explore uncertain future changes, the potential implications of change and the responses to these.

Strategy, adaptive: A defined set of responses and potential adaptations for maintaining a required performance.

CHAPTER 1
Introduction

1.1 Introduction

Bangkok is one of the World's most vulnerable megacities to flooding. The big flood in 2011 has made clear that interventions at various spatial levels will be required to make the city more flood resilient. At the grassroots level, communities in Bangkok have been very resilient over the past centuries. Many of those communities are well organized and are engaged in local flood mitigation initiatives. However, this local capacity is slowly disappearing and lacks alignment with interventions taken at city governmental level. It is becoming increasingly evident that interventions to reduce flood risk in Bangkok should be part of a holistic and adaptive strategy in order to deal with future uncertainties such as climate change and economic development. At the same time, Bangkok is a thriving, dynamic city with a huge potential to adapt to changing conditions. However, at present date there is a lack of insight in this potential of the city. This insight is vital to develop an adaptation strategy which combines climate change with socio-economic drivers.

This dissertation aims to get a deeper understanding of the Adaptation Potential (AP) of the city of Bangkok. The AP in this research is directly related to the process of urban renewal as urban renewal provides an opportunity to intervene in the urban fabric, to restore old mistakes and to make adjustments and adapt the buildings and infrastructure to changing flood conditions. This type of intervention is referred to in this dissertation as *Opportunistic Adaptation* (OA).

The main objective of this dissertation is to explore and further deepen our understanding of OA using renewal cycles of the urban fabric components for the context of Bangkok. In addition, it aims to develop a method which enables to identify and seize the opportunities of a city to adapt to an increasing flood risk using OA.

Three chapters of this dissertation (namely CHAPTER 2, 3 and 4) encompass material of published articles by the author. Their content of these chapters largely overlap with the corresponding articles, but some sections have been modified for aligning it's content with the other chapters.

1.2 Frequency and impacts of floods

Predicting future climate change is a challenge, as it is by its nature uncertain (Zandvoort et al., 2017). Implementing large scale measures and upgrading existing infrastructure to manage flood risk should be ideally a continuous process and be based on the latest insights with regards to potential future impacts caused by climate changes. The frequency and the overall impact (and changes therein) of flood events are the two driving factors for governments to take decisions regarding the potential investment in flood management measures. Flood risks can be divided into two main categories which set the boundaries of a continuum: low frequency/high impact and high frequency/low impact. Most of the time senior governments focus their attention on the first category, often resulting in responses involving large scale interventions (Botzen and Van Den Bergh, 2008). The 2nd category is more hidden, but in terms of accumulated yearly damage highly significant. The latter category calls for more localized, small scale interventions. Both require a dedicated approach, but synergies are there (Wynes and Nicholas, 2017) and a long-term strategy is needed in which the objectives of the two approaches are aligned (Zevenbergen, 2013).

In 2017, the Disaster Risk Financing and Insurance Program (DRFIP) and the World Bank (Confederation, 2016; Campillo et al., 2017) developed a risk-layering approach, which involves the development of a financing scheme focused on rapid response financing instruments, to protect against natural hazard events of different frequencies and severities. Fig 1-1 provides a graphic representation of the risk-layering approach, which is presented as a three-tiered sovereign risk-layering strategy for the national government of the G20 member countries (World

Bank Group, 2017). Combining financing instruments also enables governments to take into account the evolving needs for funds dedicated to emergency response and to long-term reconstruction. For example, in the case of an event that represents a hazard of high frequency but low severity, there are no incentives for implementing new or upgrading existing large-scale measures. This is because the impact on the community and infrastructure of such a single event is low. But aggregating the impacts of all these (frequent) small events can be substantial and even outweigh those of low frequency/high impact events on a yearly basis (Jonkman et al., 2005). The largest part of the investment needs relates to low frequency and high severity events which have the biggest impact, like for example extreme events, as shown in Fig 1-1. Although these investments ultimately benefit more frequent low impact events, they are not necessarily focused on similar disaster risks. For instance, elevating river banks to limit riverine flooding of urban areas might impair outflow of urban drainage in those urban areas during moderate rainfall events and consequently increase local nuisance flooding. The question arises of how to efficiently plan and manage areas which are exposed to high frequency events with low impact? And which governance arrangements would fit best? The most important and biggest challenges in Flood Risk Management (FRM) are related to good planning as well as to good execution of the plan in question (implementation).

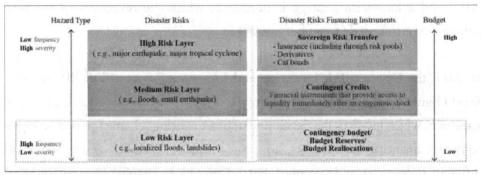

Fig 1-1. Three-tiers layering approach: Contingency budget is ideal for covering localized disasters which are a challenge for FRM

1.3 Urban adaptation

Today's cities are increasingly affected by natural disasters, of which 85% are flood related (OFDA/CRED international Disaster Database, 2013). Of these, 65% are fluvial, 10% are due to tropical cyclones and 5% are pluvial flooding influenced by monsoons. Many of the associated flood impacts are expected to be exacerbated by climate change, which is often further boosted by land subsidence (Hallegatte et al., 2013). Zope et al. (2016) assessed the flood frequency change over time of megacities such as Mumbai and Bangkok, including fluvial flooding, pluvial flooding, and coastal flooding. There has been an increasing number of flood events affecting urban areas in the last decades. This increase is largely attributed to the topographical layout of cities and their growth (i.e. urban sprawl) (Veerbeek, 2017). Urban areas are vulnerable to flooding due to poor water management and urban planning and a lack of integration between the two. Based on empirical evidence, the impact of extreme climate events calls for adapting of infrastructure and spatial layout of built-up areas to reduce economic losses. This particularly holds true for developing countries due to the existence of an adaptation deficit (e.g., Fankhausera et al., 2013). Often referred to as 'grey infrastructure', traditional large-scale interventions result from a top-down decision-making process. These include amongst others improvements in flood defenses and urban drainage systems. These top-down approaches have proved to be effective in reducing flood risk and thus limiting damage from floods occurring in cities. However, measures pertaining to this category have long lead times and require large investments. Many cities around the globe have poorly maintained infrastructure and lack the capacity (incl. the financial resources) to upgrade or replace large scale flood management infrastructure systems. Bottom-up approaches involving small scale interventions (Mataki et al., 2007; Sayers et al., 2012; Metcalfe et al., 2017) may fill this gap and serve 'to buy time' as they have short lead times and require relatively low investments. Moreover, a blended top-down and bottom-up approach may also provide an appropriate answer to effectively deal with more frequent, low impact events. Such events are typically overlooked or lack the attention of national or local governments due to their creeping nature. Bangkok is one location where such an approach is likely required, as it is prone to flooding due to urban sprawl and low-lying flat areas (Hara et al., 2005; Sintusingha, 2006; Cao et al., 2019). In addition, the implementation of large-scale interventions, such as flood embankments and retention ponds as a response to the catastrophic 2011 flood, are still lagging behind (Vercruysse et al., 2019; Laeni et al., 2019). Public spaces and public utility buildings such as local markets, social housing, and parks have been built in flood-prone areas. The economic damage caused by flooding is expected to increase to about 100 million US dollars per year in Thailand (Terdpaopong et al., 2018). There is a growing awareness that both local scale as well as collective large-scale adaptation interventions are needed to manage the increasing flood risk in Bangkok. This notion extends to all cities that are dealing with pluvial, fluvial or other types of flooding.

Apart from climate change, there are other factors contributing to an increase of flood impacts. These are related to the location, composition and distribution of the built-up areas in urban agglomerations and thus to urban planning. Planning practices of urban developments and redevelopments often ignore flood hazards and therefore are responsible for a substantial increase in flood risk which is likely to exceed the expected impacts of climate change (Storch and Downes, 2011). Furthermore, governments tend to have a 'silo orientation' meaning they often consider existing problems from a single dimension leading to responses which are suboptimal. A typical domain where such a multiple perspective is needed is disaster management, requiring the involvement of multiple disciplines, governance levels and spatial scales (Radhakrishnan et al., 2017; Pathirana et al., 2017). Pluvial flooding is often perceived as a result of an insufficient discharge capacity provided by the drainage system, which leads to the single conclusion that its capacity should be increased (Yin and Li, 2001; Miller, 2015; Wei et al., 2019).

Localized flooding caused by poor public management and inadequate stormwater infrastructure has a significant negative impact on the local residents. It encourages autonomous adaptation initiated and implemented by individuals, communities and private enterprises affected by the flood (Forsyth and Evans, 2013; Mycoo, 2014). Unclarity about roles and responsibilities among public stakeholders lead to inaction to proactively manage floods and drives local actors such as the citizens and local communities to take action and put pressure on the local government (Shen, 2007; Montgomery, 2008; Huong and Pathirana, 2013; Francesch-Huidobro et al., 2017). Since flood protection measures have been implemented as a response to the 2011 flood in a haphazard way and at some locations even have not yet been fully installed, citizens and local businesses have taken action by themselves (Lebel et al., 2011; Marks, 2015; Singkran and Kandasamy, 2016). These individual actions, referred to as autonomous adaptation in this study include amongst others the installation of floodwalls around buildings. These interventions may adversely affect the effectiveness of the existing flood management system of the city government. For example, Limthognsakul (2017), revealed from her study in Bangkok that autonomous adaptation may spread flood risk to others, especially to vulnerable groups, and thus can be maladaptive.

A city is not static, but changes over time. These dynamics encompass a continuous process of development and redevelopment of buildings, infrastructure and other urban functions. These dynamics are creating 'windows of opportunities' for flood adaptation. In turn, if these opportunities are confined to a certain area and occur within a certain time period, together they represent the AP for that specific area and time period. Some scholars conceive urban flood adaptation as a continuous, bottom-up, socio-economic process rather than as a set of isolated, stand-alone adjustments (Gersonius, 2012). Hence, a more dynamic view on adaptation seems appropriate if climate change and socio-economic drivers are considered together. The combined approach has been referred to as 'adaptation mainstreaming' (Huq and Reid, 2004; Gersonius et al., 2012; Radhakrishnan et al., 2018). This includes climate change adaptation measures that are applied and integrated into autonomous urban development and redevelopment policies, plans and/or projects. The main intention is to adapt these policies, plans and projects in order to anticipate future changes. Adaptation mainstreaming requires an integrated approach and an understanding of the effectiveness of adaptation interventions such as the value added of adjusting (components of) the urban fabric to changing climatic and/or socio-economic conditions. There are many methods and frameworks available to develop and evaluate climate adaptation strategies and policies under uncertainty. Examples of such methods are: Adaptation Mainstreaming (Schipper et al., 2003), Real in Options (RIO) (e.g., Woodward et al., 2014), Adaptation Tipping Points (ATP) (Kwadijk et al., 2010), and Adaptation Pathways (Haasnoot et al., 2013). They all increase the flexibility of the resulting strategy by allowing to change over time in response to how the future unfolds. Computer simulation tools customized for designing and decision-making purposes are increasingly used to assess flood risk and how it evolves over time in urban areas. The flexibility of different risk management approaches is evaluated and compared using these simulation tools by identifying various responses to be implemented over time and their effectiveness by assigning a utility value to these responses (Fazey loan et al., 2016, Gersonius, 2012; Revi et al., 2014). The practice of adaptation mainstreaming is not new. For example, Schipper et al. (2003) provided historical evidence that the adjustment of standards for building codes and flood management infrastructure in North America largely reflected the observed past trend of climate change.

A mainstreaming adaptation approach is achieved through aligning processes and actions in different fields, such as water management, architecture, landscape and urban design, creating additional benefits, for instance to the local community (Masoud et al., 2014). Some well-known examples resulting from this approach include: water squares, urban wetlands, and amphibious architecture (Van der Pol, 2011; Anderson, 2014; Evans, 2015). However, these mainstream adaptation approaches do not provide guidance on how to select the location and the flood protection measures to be implemented. Methods such as RIO and ATP are used to address the timing of climate change adaptation actions.

RIO is founded on the basis of 'options theory', which considers uncertainty as an opportunity and identifies flexibility in creating and exercising an intervention. It encourages the exploration of all possible interventions in an uncertain decision environment and values the interventions based on net present value and flexibility (Gersonius and Zevenbergen, 2013). RIO is a technique which can be used in urban water engineering to incorporate and value flexibility of 'engineered water systems' (Radakrishnan et. al., 2013; 2017). It focuses on creating flexible alternative designs and compares the cost of alternative designs based on the optimal economic efficiency. However, RIO does not provide a plausible sequence or sequences of measures (pathway(s)) to be implemented over time. It can potentially be combined with a bottom-up approach and be useful in evaluating the performance of an adaptation strategy under uncertainty in monetary terms (Wang and De Neufville, 2004; Gersonius, 2010; 2012; Woodward et al., 2011; 2014; Means et al., 2010).

Adaptation Pathways involve a structured approach for designing climate adaptation strategies and policies based on potential sequences of adaptation options to be implemented over time and to ease adaptation in the future. ATPs provide the conditions at which an adaptation option (of an adaptation pathway) no longer meets the performance requirements (Haasnoot et al., 2011). Adaptation Pathways are about identifying potential options to provide flexibility, which allow the system to adapt to climate change. However, in its simplest form, it lacks a clear procedure for the development of an "optimal" dynamic adaptive strategy. Both methods do not provide a procedure for combining with the bottom-up approach, but it can deal with changing conditions by transforming them into opportunities.

Despite the contribution of the abovementioned methods to strategy and policy development, they are not designed to identify and exploit adaptation opportunities arisen from 'autonomous' renewal and maintenance processes. Moreover, most examples of their application (using a concrete context) published in the literature reveal that they comprise only a limited number of predetermined adaptation options and pathways. This is due to practical constraints following from the requirement that outcomes of the methods should be visualized in a human-readable format (Forsyth and Evans, 2013; Mycoo, 2014; Kwakkel et al., 2015). In particular there is a need for tools to effectively solve architectural or urbanism related issues that need to be handled automatically by software due to their high degree of complexity.

The exploitation of autonomous redevelopment cycles for climate change adaptation is conceptualized in this dissertation as "Opportunistic Adaptation". OA requires an integration of climate adaptation into asset management planning of urban infrastructure and renewal schemes of buildings and public spaces (Alegre et al., 2012). The spatial dimension and nature of OA interventions vary from small scale upgrading of single objects (such as an individual building or a small public square), to large-scale transformations such as an integrated design-based redevelopment project comprising an entire neighborhood once the area requires major restructuring. Examples of interventions to reduce flood risk include the integration of a green roof in a refurbished building, a water square (i.e. urban water detention area) replacing a traditional town square and an elevation of a complete neighborhood by raising the existing buildings. Each of these local interventions impacts the flood risk of a city but some may become maladaptive when this risk is spread to other areas (Linithongsakul, 2017). These local interventions can be combined with other neighboring interventions to further reduce the flood risk. Hence, OA needs to be managed in a concerted way. Only then a city will have the ability to fully exploit its AP to reduce flood risk and seize the adaptation opportunities synergistically.

1.4 Adaptation potential and capacity

The lifespan of individual assets defines the potential for an adjustment over a longer time period. Similarly, adaptation of an urban neighborhood or city involves an iterative process in which the lifespan and the lifecycle of assets drive a recurrent adaptation process that continues indefinitely in the future. Depending on the type and scale of the interventions, they can become significant factors in molding the form and space of the urban fabric over time.

It follows from the above that an opportunistic and synergistic approach in urban planning in which the 'urban dynamics' are driving the Adaptation Rate (AR), provides a process of continuous adaptation. While the potentials for OA have been identified in several papers (e.g., Zevenbergen et al., 2008; and Veerbeek et al., 2010) many aspects related to this new approach are in their infancy. For instance, the effects of differentiation in age and lifespan of urban components (e.g., different types of buildings, streets, etc.) as well as the portfolio and scale of measures that can be integrated during redevelopment are hardly understood. This hampers the appraisal, effectiveness and ultimately the integration into actual adaptation strategies. Furthermore, an assessment of the method in different case study areas with different hazard as well as urban characteristics has not been performed. OA, therefore, remains currently in the realm of academic research. To assess the AP, the adaptation opportunities have to be defined. The first step is to gain insight into the lifespan of urban assets (a proxy for the adaptive potential) and to develop a method to identify and map individual assets which are reaching the end of lifecycle (EOLC) at a similar future moment in time. Based on this information clusters can be formed of multiple assets which collectively are reaching their EOLC. In other words, a method needs to be developed which allows for the assessment of the spatiotemporal distribution of urban renewal cycles, and which identifies clusters of assets which are reaching the same EOLC. This tool will be pivotal in strategic master planning which targets to integrate future adaptation measures into an urban transformation process that gradually adapt existing cities to the changing needs and circumstances. This research aims on the one hand to provide a better scientific understanding of the potential for OA, and to the other to deliver a tool to be applied by practitioners on 'when and how to adapt'.

1.5 Research objectives

The goal of this research is to explore and further deepen our understanding of OA using renewal cycles of the urban fabric components. This is done by a combination of activities comprising an assessment of the AR and potentials for interventions (implementing adaptation measures) in the urban fabric at various spatial scales (individual building, block and neighborhood level) using model simulations and a dynamic design approach. In addition, it aims to develop a method which enables to identify and seize the opportunities of a city to adapt to an increasing flood risk using OA.

1.6 Research questions

In the following section, the research questions regarding the analysis of the potential of OA are given. These research questions are:

How can Opportunistic Adaptation be operationalized and applied in an actual urban area in support of Flood Risk Management? How to assess the Adaptation Rate(s), and associated adaptation options? And what are the governance arrangements required to better facilitate Opportunistic Adaptation?

The main research questions can be subdivided into four sub-research questions:

1. What are the components of an operational method to identify adaptation opportunities based on autonomous urban renewal cycles?
2. Which factors influence the Adaptation Rate and capacity and how do they relate to the spatial dimension of a city?
3. How does Opportunistic Adaptation affect the flexibility of adaptation options over space and time?
4. What institutional arrangements does Opportunistic Adaptation require to flourish given the different stakeholders involved in urban climate adaptation?

OA calls for strategic planning and long-term thinking. To answer these research questions, criteria and constraints underlying strategic planning and long-term thinking have to be identified to narrow the scope of this research and to develop a research framework of practical relevance.

1.7 Relevance and applicability

The outcomes of this research aim to identify the potential benefits which can be gained from OA. These benefits are dependent on the end-user demands. Two types of end-users can be defined: i) the citizens (urban communities and social benefits) and ii) different professionals and institutions, such as government, water management institutions, developers, urban planners, etc. They require accurate, reliable and readable information to estimate the time needed to take action in retrofitting or adjusting buildings and/or urban infrastructure for the purpose of preventing, preparing and protecting against flood events. Hence, the anticipated benefits of the research outcomes are that they will support decision in FRM and raise awareness among people, community and government. This awareness will increase their understanding of how to mitigate the adverse impacts of flood events. In addition, the research also contributes to a more practical approach for integrating FRM into urban development. Optional flood adaptation measures (FAM) offer more flexibility as they increase the degree of freedom with which flood solutions can be selected.

Adaptation strategies based on exploiting opportunities provided by urban regeneration are still very academic and thus theoretical. Practical applications are lacking. This research aims on the one hand to provide a deeper 'theoretical' understanding of the potentials of OA, and to the other hand to deliver a practical tool for applying this method. Hopefully, the Thai government can use the outcomes for accessing policies and projects that can help people in the future and use less money than stand-alone measures.

1.8 Action research

This research aims to pioneer an advanced method in design and computation. It is based on five years of research following an interdisciplinary, case-study focused approach to design fostering collaboration with end-users. It examines architecture and urbanism through the lens of climate adaptation. It seeks to reach out to the next generation of flood managements and architects who will likely take a facilitating role.

I actively participated in the case study by organizing and moderating two workshops in the case study area of this research. Participants of these workshops included technicians, decision-makers, and practitioners from local government agencies.

The action research in the study consists of two threads. The first thread is about developing a framework. Once the theoretical framework model has been set-up, it needs to be translated into a methodology. The second thread is concerned with managing the operational method. The case study in the Lad Krabang district of Bangkok has been conducted to evaluate the technical feasibility of the OA approach in practice in order to address research questions 2, 3, and 4. The case study was also used to gain insights into the mechanisms that determine the adoption and mainstreaming of the OA approach (research question 4).

The operational method that was applied in the case study can be considered as Participatory Action Research (PAR) (Graaf, 2009). Traditional research aims to advance knowledge by developing theories and testing hypotheses. PAR is not (only) to advance scientific knowledge, but to achieve and improve practical objectives. Therefore, to evaluate the research outcomes (model & tools), stakeholders & potential users have been engaged in interactive workshops, which have been initiated by the author, and in customized surveys to conduct an (*ex-ante*) evaluation.

Finally, interviews have been held with representatives of the national and local governments to validate these findings. For example, the reports capturing the discussion and conclusions of the workshops have been shared with the Bangkok Metropolitan Administration (BMA) and after receiving and processing their feedback, the reports have been published and used to further improve the OA approach.

It follows from the above that in this dissertation, the results of the case study have been used to demonstrate, improve and test the OA approach.

1.9 Dissertation outline

This section provides a description of the chapters of the dissertation. They comprise of:

- *Introduction.* CHAPTER 1 contains a general introduction of the study and briefly explains the background and the rational of this research. It also reviews the state-of-the-art of relevant research including the gaps in knowledge that exist and needs to be addressed.

- *Towards an operational method.* CHAPTER 2 reviews the relevant literature on (climate) adaptation and provides basic background information on the OA approach. Based on this review this chapter also presents the components of an operational method for OA. It introduces the basic features of the method, the framework, as well as the models and evaluation methodology required to address the research questions. Finally, this chapter describes the criteria for the selection of the case study area.

- *Assessing the adaptation potential and opportunities of an urban area.* CHAPTER 3 elaborates deeper on the concept of AP and opportunities of an urban area and their relevance in the domain of climate adaptation. The key components of an adaptation opportunity, which can be detected and extracted from an assessment of the lifecycle, are explored. The lifecycle of buildings and infrastructure of a defined urban area in Bangkok is analyzed and presented in this chapter.

- *Applying FAM at different spatial levels in Kehanakorn.* CHAPTER 4 presents decision tree diagrams and potential FAM (e.g., store, delay, resist, and avoid) to be implemented over time (depending on the lifespan of measures), which are referred to as *adaptation pathways* in this study. These pathways have been used to create a sequence of dynamic designs to visualize the transformation process and to explore potential adaptation strategies.

- *Decision tree method for evaluating the flexibility of flood risk adaptation options.* CHAPTER 5 ference source not found. describes the development and application of a novel methodology for evaluating the flexibility of adaptation options (set of measures). The methodology explores the use of decision tree diagrams to apply different metrics.

- *Challenges of implementing an Opportunistic Adaptation approach in Bangkok.* CHAPTER 6 addresses the governance aspects required to implement the OA approach. The Kehanakorn village 2, Lad Krabang district, is used as a case study. Particular attention has been paid to the enabling conditions (such as required knowledge and awareness).

- *Conclusions and recommendations.* Finally, in CHAPTER 7, the outcomes of the different chapters are summarized and related to the initial research questions. The hypotheses are evaluated. This chapter completes with recommendations for practice and for further scientific research to advance and develop the OA approach.

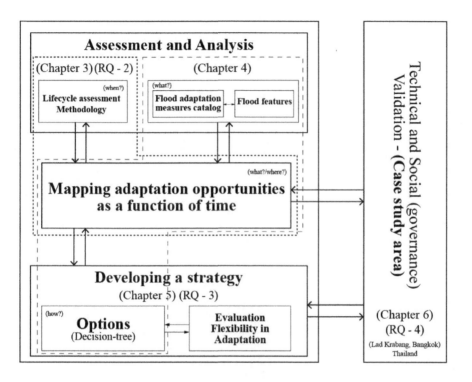

Fig 1-2. Outline of the dissertation

CHAPTER 2
Towards an Operational Method

Operational method	
	CHAPTER 2

Assessment and analysis

CHAPTER 3		CHAPTER 4
Assess adaptation potential	Identify adaptation options	Assess current and future flood risk
lifecycle assessment methodology	Catalogue and classification	Flood modelling empirical data (beyond scope of this study)

Developing adaptation pathways

CHAPTER 3 + CHAPTER 4	CHAPTER 3 + CHAPTER 4	CHAPTER 4
Define clusters	Map opportunities	Identify pathways
Urban density, synergies co-benefits	Selection of options	Decision Tree method

Towards a strategy

CHAPTER 5	CHAPTER 6
Evaluating flexibility	Governance
FIA method (measuring flexibility)	Interviews and survey. Ex ante evaluation

This chapter is based on the journal article "Nilubon, P., Veerbeek, W., & Zevenbergen, C. (2016). Amphibious architecture and design: a catalyst of Opportunistic Adaptation? – case study Bangkok. *Social and Behavioral Sciences*, 216, 470-480.

2.1 Opportunistic Adaptation

An approach that involves OA requires a systematic perspective, in which the city is perceived as a collection of interacting components (e.g., buildings, road, and etc.), constructed at different moments in time and with different lifespans. Depending on the lifecycle management and replacement strategy of each individual asset, assets are renewed individually or as groups (e.g., a complete street including buildings, pavement, gardens, sewage network, and other utilities). Thus, from a theoretical perspective, the complete renewal of a given area depends on the most recently constructed assets with the longest lifespan, i.e. the last components to be replaced. The distribution of assets reaching the end of life over a given range of years defines the actual replacement rate. In OA, as the upgrading of assets is performed during replacement, the AR of an area coincides with the replacement rate of the individual assets.

Due to its orientation on assets, OA enables the development of a decentralized adaptation strategy. Asset renewal starts at the individual building level, but may provide an opportunity to adjust the design (and even the function) of an urban neighborhood to new insights, standards or policies for instance driven by projected changes in the intensity and/or frequency of a natural hazard, such as a flood. This warrants a strategic, concerted, and locally driven approach often catalyzed by a recent hazard experience and/or by information provided by experts on the risk including the sensitivity and exposure of the assets.

It follows from the above that a methodology to assess OA will require information on:

1. The spatio-temporal distribution of assets in an urban area, including the lifespan of different assets, their spatial distribution, age and end of lifecycle (EOLC);
2. Potential local adaptation measures;
3. Evaluation criteria. The metrics used to measure the performance and

resulting effectiveness of OA as a function of time;

4. Scenarios. Future adaptation scenarios as a function of replacement strategies and application of measures at different scale levels.

2.1.1 Spatio-temporal distribution of urban areas and timing

In urban planning, design and architecture, the construction period of assets is one of the determining factors that defines how urban areas look and more importantly to what extent they function under the current conditions. In many cities, traversing from urban center to outskirts means traveling from the oldest parts to the newly constructed suburbs. This is especially the case in many European cities which revolve around a historical center. Yet, over the years many micro and macro scale redevelopment cycles have taken place, creating a large differentiation in the construction age of individual assets. Only in truly historic centers, as for instance in many Italian cities, clusters of historic buildings with similar construction dates can be found. On the other hand, rapid urban expansion, as for instance in post-war Europe or in contemporary Asian cities, produced large clusters of buildings of identical age. Planning models for cities differ significantly, varying between classic concentric revolving around a single CBD and network cities consisting of many (multi-functional) clusters. Furthermore, a large proportion of urban areas have been developed without any top-down planning regime. UN-Habitat (2013) estimates that in developing countries 95% of the current urban development is unplanned (Jones et al., 2015). Although this does not necessarily create a large

differentiation in spatio-temporal distribution of the urban building stock, it does mean that schematization of cities from planning documents is often impossible or at best partial. Finally, planning models are applied differently depending on the physical constraints (e.g., the shape of a delta), resource distribution (e.g., water resources), as well as many socio-economic, cultural and political factors. For instance, the shape of the coastline or the morphology of mountainous areas with steep slopes will largely determine the suitability to develop built-up areas. Similarly, the tradition of living in detached housing will create a very different distribution of built-up areas compared to areas where living in large apartment flats is the norm. For instance, the decision to exclude an area for occupation in order to develop a 'green belt' will change both the urban shape and the composition of individual neighborhoods, streets, and blocks.

The initial construction period defines the spatiotemporal layout of cities. The actual lifespan is important to be able to assess when incremental upgrades and replacements have taken place. For instance, considering that a street has been initially constructed 70 years ago and has undergone major refurbishments on average every 30 years, it is expected that the last refurbishment took place approximately 10 years ago. The asset, therefore, reached a third of its lifespan and is estimated to undergo major refurbishment or replacement after 20 years (see Fig 2-1).

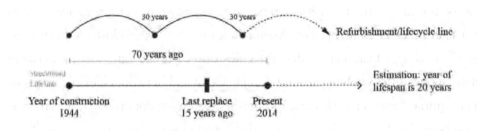

Fig 2-1. Diagram of estimation of year of lifespan (e.g., street/road)

Obviously, the actual moment of refurbishment or replacement is determined by the maintenance and depreciation regime. Road maintenance, upgrading, and replacement can, for instance, be i) performed as part of a predetermined sequence based on the estimated lifespan of the different road types, ii) be based on continuous monitoring and application of performance criteria or iii) be incident based, which means that refurbishment is based on reported failure or malfunctioning. Houses could easily last for over a century, although for instance in Japan the average lifespan of a detached house is only a few decades since both the construction method as well as the depreciation scheme are based on a relatively short lifespan (Banaitiene, Banaitis et al., 2008). The lifespan of underground sewer and drainage structures is often stretched beyond the theoretical; in London for instance the age of sewage and drainage structures often exceeds 150 years (Cotterill et al., 2020). Lack of financial resources prevents actual refurbishment or replacement assets and their performance often drops below the required standard (Cole and Kernan, 1996). Furthermore, refurbishment and replacement of assets is often not performed on an individual basis. For instance, refurbishing of sewer and drainage pipes is preferentially combined with a refurbishment of the street, sidewalk and other related functions above ground. In turn, redevelopment of multiple adjacent buildings or blocks is sometimes combined with street refurbishing. The resulting distribution of the EOLC for all assets in a given area could, therefore, vary significantly. Recently constructed buildings that only reach the EOLC in 80 years could be adjacent to older structures that are fully deprecated. Over the last decades, asynchronous replacement of assets can evolve following a complete refurbishment or replacement of larger clusters. This complex behavior in space and time defines the actual urban dynamics used in OA to mainstream upgrades and adaptation measures into the building and urban infrastructure. This also means that the resulting AR may differ significantly between assets, clusters, neighborhoods and ultimately between cities.

Over the last decades, the planned lifespan of buildings and infrastructure has decreased. The fact that current building practices are based upon the implicit assumption that the built environment will not experience significant change, will have major ramifications for investment and building methods. Veerbeek and Zevenbergen (2009) have assessed the rate of substitution of buildings and other built structures to identify opportunities to integrate floodproofing schemes into urban areas of some European cities. In this study two aspects of timing have been considered: year of construction and the EOLC.

2.1.2 Uncertainty and planning horizon

Climate adaptation calls for long-term planning. But, while focusing on this long-term horizon, constraints and opportunities for their implementation which arise from short-term measures, should be identified and taken into account as well. While the use of climate change scenarios to identify future hazards has become common practice, the development and application of long-term socio-economic scenarios in climate adaptation are still rare. One of the few studies in the field of FRM that explicitly uses long-term socioeconomic scenarios is the OST Foresight Report: A Future analysis of UK coastal flooding and erosion (2005). This study involves a national assessment of the changes in future flood risk and effectiveness of responses for a range of climate change and socio-economic scenarios in the UK. The use of scenarios to develop climate adaptation strategies often implies assigning a likelihood to the potential consequences of the scenario by using a probability distribution. Alternatively, scenarios are used to identify the ranges of future hazard levels. Robust decision-making, no regret planning and/or various methods to incorporate uncertainty into planning (e.g., Mens et al., 2013) are subsequently used to adopt a long-term horizon into an adaptation strategy. While many strategies incorporate long-term climate change projections in defining design standards and in selecting and designing subsequent measures (infrastructure and other components of the urban fabric), the operational lifespan and potential future upgrading are often only implicitly incorporated into the actual design and management of the implemented measures. Furthermore, many stand-alone measures create a 'lock-in'; the chosen adaptation strategy and subsequent adaptation pathway cannot be altered in the future (Walker et al., 2001, Haasnoot et al., 2013). Often such constraints occur if it concerns large scale structural measures, e.g., a dike system protecting the hinterland against river flooding.

Instead of being designed to protect an area as a whole, OA-based FAM are usually integrated at the level of individual assets, with floodproofing measures applied at the level of individual buildings, streets or blocks. This makes the whole OA process decentralized, since the involved measures are being applied gradually, as each individual asset reaches its End of Life-Cycle (EOLC). "Kokx and Spit (2012)" refer to the unbanked areas of the Feijenoord neighborhood in Rotterdam, where protection standards require new houses to be elevated one meter above most of the current street level. A variety of elevations and housing typologies are present in the area, which represent a relatively accurate reflection of previously accepted standards (HSRR3.1, 2013).

2.1.3 Urban components

The term 'agile architecture' refers to the evolutionary design of a system which allows to change over time, accommodating the needs of the current users. It is increasingly being recognized as a requirement of a dynamic society, the existence of an alongside and ongoing design process producing a constantly changing built environment (Brand et al., 1994; Prins, Bax, Carp et al., 1990-1993). The functional requirements of urban architecture are changing, as well as other factors that alter building preferences, market conditions and technical constraints that determine the valuation of buildings as well as other assets. In the past, this often implied the development of flexible building concepts. For instance, the 'open building concept' in which the actual layout of the floors can be changed according to the occupants' requirements, has become the dominant model in the office market. One of the fundamentals of the 'open building concept' is the principle of distinct levels of intervention: a building can be conceived as a collection of several layers of longevity of built components. Parts of a building can be removed and replaced, such as the façade of a building, revealing a layer that is independent of its structure. Duffy (1990) identifies a perspective of buildings in which the building consists of different shearing layers: shell or structure layer, services layer, scenery or layout layer, and 'mobilia' layer (i.e. furniture, appliances and other movable accessories including curtains and rugs that make a home livable). From an economic point of view Templemans (1990) and Templemans et al. (2002) describe a building as an aggregation of stock services with different life spans.

This perspective can be expanded to other asset types. For instance, streets can be decomposed into different lanes accommodating different kinds of traffic (e.g., car lanes, bicycle lanes, and sidewalks) as well as swales and other adjacent zones. Depending on the materialization and detailing of the streets, tiled sidewalks can easily be repaired, refurbished or replaced at local level or along the complete street at any given moment in time independent from the rest of the street profile. Asphalt roads that include several layers of underlay (used for stabilization and for strength) might require more comprehensive repairing, refurbishment and ultimately, replacement strategies. These, in turn, might include manholes and underground utility pipes and lines. In other words, as for buildings, roads can be seen as inclusive structures consisting of multiple (sometimes interlinking) components.

A logical extension of this perspective is the notion of scale. While assets can be decomposed into different components, the perspective can be changed in a similar fashion to block, neighborhood or even city level. In this case a block could include the private gardens inside the block as well as the adjacent streets, parking lots and other functions in close proximity. The applied scale, therefore, determines the level of detail in which components are considered.

Decomposition of assets, or groups of assets into different components, can be performed from different perspectives. While previous examples are based on an object-oriented view, an alternative perspective could be applied that focuses on urban metabolic flows (e.g., energy, waste), material properties or any other classification scheme.

2.2 Local adaptation measures

Standards and measures to adapt structures in managing natural hazards come in many shapes and sizes. For instance, Pötz et al. (2012) published a comprehensive catalog of urban adaptation measures focusing on a multitude of topics including heat stress, urban floods, water quality, etc. An important characteristic of a measure is the protection level it will ensure. The protection level is derived from a comprehensive probability distribution of events (e.g., intensity-duration-frequency curves in case of rainfall events) or based on historic events from which the actual return period is unclear (e.g., a water stage in a complex delta system). The context of the urban complex does not mean that measures can be integrated to comply with a given single design standard. It could be that there are functional, technical, cultural or socio-economic constraints which might cause specific measures to comply to the design standard only to a given degree. For instance, elevating buildings to withstand floods by putting them on columns might not coincide with the desire to provide a pedestrian friendly, vibrant downtown area at ground level. Similarly, integrating 'bioswales' into a street profile might not fit within the available space to accommodate cars, bicycles and pedestrians in different lanes. On the other hand, measures can be scaled up at moments in which multiple assets, which are located in close proximity to each other, are being refurbished or replaced. For instance, using a single private garden to infiltrate rainwater from a disconnected rooftop might have some local effect. Yet, when gardens of several houses are combined into a single peak storage facility for rainwater, the combined effect is likely larger than the sum of the constituent elements. Furthermore, scaling up a measure might offer opportunities for integrating additional measures: trees to reduce heat stress, a small wetland used to clean stormwater as well as other measures that cope with multiple pressures and hazards. Finally, the combined plots can be used to increase the overall livability of the area, providing a semi-private park for the residents including for instance a children's playground. Hence, in this example, scaling up adaptation measures provides multiple benefits: it achieves a further reduction of the flood impacts and improves the livability (probably at lower costs compared to the application of stand-alone measures that only provides single functionalities).

2.2.1 Co-benefits of flood adaptation measures

FAM can provide two types of co-benefits: 1) benefits at a local level, which are generated by a single, localized intervention, and 2) benefits resulting from the upscaling of multiple interventions that through synergy create an impact on an area. For example, the Water square at the Benthemplein in the city of Rotterdam, gives communities and local stakeholders (students) the opportunity to use this space for recreational and social activities, while improving the environmental conditions (de Graaf and der Brugge, 2010). This measure is therefore providing benefits at a local level. The upscaling of flood adaptation measures can potentially deliver benefits by connecting individual measures into larger networks. This is for instance the case when patches of blue-green infrastructure (e.g. bio-swales, raingardens, etc.) are connected and provide ecological corridors. Typically, flood adaptation measures provide opportunities to address other challenges as well such as purifying polluted water, improving air quality, enhancing local amenity value, etc. Upscaling adaptation measures usually enhances these potentials. This is particularly true for green infrastructure. Green as well as blue-green infrastructure are defined as 'an interconnected network of green space and surface water that conserves natural ecosystem values and functions and provides associated benefits to human populations' (Benedict and McMahon, 2002; Benedict and McMahon, 2006). There is growing evidence that connecting the fragmented urban green spaces of a city into a vast network will amplify the co-benefits of green infrastructure (Gill, Handley et al., 2007). This holds true for both the quantity (the magnitude of a single benefit) as well as for its quality (the number of co-benefits). These networks are known for enriching habitat and biodiversity, cleaning air and water, increasing recreational opportunities, improving health and better connecting to nature and sense of place, and finally increasing property value.

2.2.2 Evaluation Criteria

Measuring the benefits of adaptation measures in urban areas often focusses on assessing the avoided impacts in monetary terms (comparing the hazard impact before and after implementation of the intervention) for a set of events with given exceedance probabilities against the costs of the intervention. Yet, to assess the flood hazard impacts covering a wide range of tangible, intangible, direct and indirect impacts need to be considered (e.g., Zevenbergen et al., 2010). They often require sophisticated frameworks, models and extensive datasets (e.g., Veerbeek et al., 2009). When covering multiple hazards (e.g., floods, drought and heat stress) a comprehensive assessment of impacts becomes almost intractable. Even for a single hazard type (e.g., pluvial flooding) impact modeling is often limited to the estimation of direct damages, neglecting for instance health impacts, traffic interruption and many other consequences that require substantial empirical data from historic events. Especially when integrating climate change projections which include unprecedented extreme events, an assessment of long-term impacts is often clouded by substantial uncertainties. Empirical methods exist that address for instance on the exposure to an event in combination with mortality functions for the estimation of loss of life (Jonkman, 2007) or loss damage curves (Kokx & Spit, 2012). Counting the number of affected people, the number or area of exposed properties or the number of blocked road segments can already provide valuable information about the severity of an event. Various post analysis methods can be used to help interpreting the outcomes, including for instance the response proportionality (De Bruijn, 2006) or simply the application of a threshold that indicates to which amplitude a hazard is acceptable (e.g., the minimum acceptable flood depth).

An important characteristic of an impact assessment of natural hazards is the level of detail at which the analysis is performed. Flood risk assessment, for instance, is often performed at a regional scale, using broad land use and land cover classification schemes as data for the application of damage curves or other impact calculation models. When working at city or even neighborhood level, the required level of detail should be of a finer resolution; individual assets (e.g., buildings, roads, etc.) should be accounted for in the applied models. Furthermore, the resolution on which the extent of the hazard is calculated (e.g., the flood extent) should coincide with that level of detail in order not to compromise the precision and power of prediction of the calculations (Werner, 2001).

Apart from the avoided flood risk due a given adaptation measure might provide, additional metrics can be included to evaluate some of the co-benefits the measure delivers. Metrics to evaluate the livability of an urban area have been developed, although these are often based on extensive surveys and stakeholder participation For example, measures that help improving the livability by increasing the fraction of pedestrian infrastructure might be one metric by which this can be evaluated. Community design is a design process based on community participation, which emphasizes the involvement of local people in the social and physical development of the environment in which they live (Toker, 2007). On the other hand, this needs to be balanced with the effort put into to increase the quality of urban water management practices, which can be evaluated by the presence or absence of certain features, such as the extent to which it is embedded into the street design. Additional factors could be used to assess the flexibility of an urban plan for an area. Here, an explicit spatiotemporal evaluation of the EOLC could provide information on when (larger) areas are available for redevelopment. This allows to identify when an area can be adapted to possible new future requirements beyond the realm of natural hazards and more focused on typical urban properties and services such as livability, functionality (i.e. land use), typology, density, etc.

2.3 The operational OA framework

Cities across the globe are facing competing priorities as their budgets are always limited. Climate change adaptation is often considered as yet "another priority". Tools for urban planners and decision makers to identify, evaluate and prioritize potential adaptive responses are becoming increasingly important (Kwadijk et al., 2010; Haasnoot et al., 2013; Buurman and Babovic, 2016). There is still a gap in the understanding of the effectiveness of adaptation actions: the costs are upfront, but the benefits are remote and uncertain. This lack of understanding complicates the identification of when and where a city has to adapt. Alternative strategies involve (i) determination of "no-regrets" actions that generate net social or economic benefits independent of climate change (Bulkeley, 2001; Smit and Pilifosova, 2003) and (ii) seizing the 'window of opportunity', adapt where the opportunity arises from autonomous maintenance, refurbishment or redevelopment activities referred to as OA. In this research, OA is defined as:

> *An adaptation strategy where adaptation measures are integrated into the design of assets or groups of assets and are implemented when existing urban assets reach the end of their lifecycle (EOLC).*

The occurrence of such opportunities is dependent on a range of factors, including functional, technical and economic depreciation schemes that determine the lifespan and subsequent renewal rate of individual and clusters of urban assets, including buildings and infrastructure.

Although many cities, particularly in Asia and Africa, are challenged by an unprecedented growth of new urban areas, climate associated impacts typically occur in the historic parts of the city. Hence, it is important to integrate adaptation into mainstream renewal activities in such a way that the existing city gradually transforms into a climate resilient city to better cope with existing and future climate challenges (Pelling, 2010). The big advantage of this gradual transformation process is that adaptation does not depend on the implementation of a single, large scale one-off measure that, although robust, might not be able to be adjusted in the future to address unforeseen challenges and uncertainties (Wilby and Dessai, 2010). In contrast, OA leads to climate adaptation measures which are more flexible and less invasive allowing for a gradual change, which is easier to apply and cope with changing future insights and standards. It allows for continuous and incremental adaptation of urban areas. The lifecycle-centered approach in OA is a bottom-up process which is not driven by the availability of external resources, political will or societal pressure from out-side. Instead, OA is giving clues "where to adapt" and circumvents the largely subjective question of traditional adaptation strategies of "when to adapt", which too often serves as an excuse for inaction. Furthermore, due to its bottom-up and inclusive nature this approach also stimulates the acknowledgement and incorporation of multiple benefits in the selection and design of adaptation measures for instance the recognition that certain measures also increase property value or boost livability or health benefits (Veerbeek and Zevenbergen, 2009). This is particularly demonstrated in small scale, blue-green infrastructure projects. (Ahern, 2013; Rozos and Makropoulos, 2013; Thorne et al., 2015; Voskamp and Van de Ven, 2015; Kati and Jari, 2016; Bottero, 2018; De Roeck et al., 2018). Better integration of climate change-adaptation into urban renewal cycles may also lead to a cost-reduction compared to stand-alone measures. Depending on the opportunities, climate adaptation might be facilitated through design (for example to systematically explore the feasibility and multi-functionality of different housing typologies) instead of 'patching-up' existing structures that might not be equipped to incorporate for instance dry-proofing measures to provide flood protection. The ultimate goal of OA is that it will at the end lead to a water sensitive city in which the provision of ecosystem services, enhanced livability as well as resilience to natural hazards are co-created and integrated into a coherent urban design (Munang et al., 2013). This requires information about the individual adaptation opportunities and of the value added of clustering of these opportunities. It therefore seems important to identify opportunities for clustered adaptation of assets based on their proximity and EOLCs. Obviously, clustering calls for a customized approach and thus differs from urban neighborhood to urban neighborhood. For example, some areas are very heterogeneous in composition and construction age which limits opportunities for upscaling measures. Yet, other areas might be more homogenous and can be retrofitted/replaced in conjunction, at a future window in time. Thus, mapping of adaptation opportunities over a larger area is a prerequisite to identify a region's potential for OA. Yet, to do this, first a method needs to be developed to estimate the EOLC of actual urban assets as well as to facilitate the clustering of adaptation opportunities in space and time.

In order to operationalize OA, one needs to understand the lifecycle aspects of assets, but also the spatial structure of the city. In order to do so, an OA method should focus on three components:

1. The EOLC of urban assets in a given area of interest;
2. The spatial differentiation in EOLC-distribution in order to identify areas where assets reach the EOLC at a similar future point in time;
3. The selection of an (effective) adaptation measure that can be integrated in the area where an asset reaches the EOLC;

Together, these components map out the spatiotemporal adaptation options for a given area and determine the rate in which the urban area can be adapted to future climate change. Obviously, the AR depends on the spatiotemporal characteristics of the individual assets. For instance, an extensive social housing area constructed in the same period (e.g., in the post-war reconstruction period in Europe) provides future opportunities for a large-scale integrated adaptation intervention. In contrast, a highly differentiated downtown mixed-use area allows for gradual adaptation at the level of an individual building or block of buildings. The final question is of course if the resulting AR is sufficient to keep up with the expected rate of climate-induced changes of extreme weather events. This assessment requires insight into the local characteristics of climate change which are typically highly uncertain. The AR can be evaluated on the basis of the performance of the interventions by comparing them against a set of prechosen projections. Based on above, the operational method is presented in Fig 2-2.

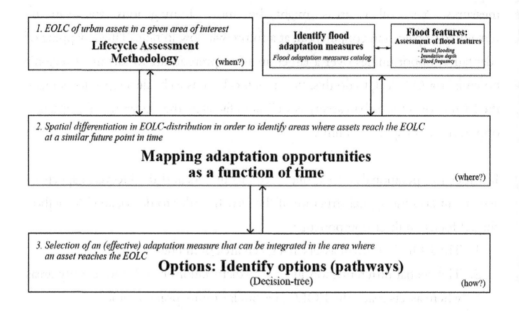

Fig 2-2. The operational OA method

CHAPTER 3
Assessing the adaptation potential and opportunities of an urban area

Operational method
CHAPTER 2

Assessment and analysis
CHAPTER 3
CHAPTER 4

| Assess adaptation potential | Identify adaptation options | Assess current and future flood risk |
| Lifecycle assessment methodology | Catalogue and visualisation | Flood modelling/empirical data (beyond scope of this study) |

Developing adaptation pathways
CHAPTER 3 · CHAPTER 4   CHAPTER 3 · CHAPTER 4   CHAPTER 4

| Define clusters | Map opportunities | Identify pathways |
| Urban density, synergies/co-benefits | Selection of options | Decision Tree method |

Towards a strategy
CHAPTER 5
CHAPTER 6

| Evaluating flexibility | Governance |
| FIA method (measuring flexibility) | Interviews and survey, Ex ante evaluation |

This chapter is based on the journal article "Nilubon, P., Veerbeek, W., and Zevenbergen, C. (2018) Integrating Climate Adaptation into Asset Management Planning: Assessing the Adaptation Potential and Opportunities of an Urban Area in Bangkok, *International Journal of Water Resources Engineering*, 4(2), 50-65."

3.1 Introduction

The current practice of urban climate adaptation is often limited to stand-alone measures to meet estimated future climate change driven demands. These interventions typically comprise of incremental upgrades of existing infrastructure systems such as the deepening of channels, increasing the pipe drainage capacity and pumping capacity. When measures concern the upgrading of existing infrastructure, they can often be regarded as adaptation actions in response to the impact of a major climate event (e.g., Copenhagen Municipality, 2012; Liu & Jensen, 2017).

The existing urban fabric of cities changes over time as buildings and infrastructure are continuously being maintained, upgraded or replaced. These urban redevelopment cycles provide opportunities to adapt these systems to future challenges imposed by climate change (Mega, 1996; Zevenbergen et al., 2008). Seizing these opportunities, requires the integration of climate adaptation into the asset management of urban systems (e.g., Alegre et al., 2012). The scale and nature of the integrated adaptation measures vary from small scale upgrading to extensive integrated design-based projects, where complete neighborhoods are redeveloped to better cope with climate extremes once they become outdated. Adaptation to more extreme rainfall events, includes examples such as the integration of green roofs into new building design, the transformation of public parking lots into water squares (urban water retention areas) as well as the elevation of complete neighborhoods to a higher level. It is important to note that these interventions rolled out across a city may substantially impact its overall performance. Hence, if managed in a concerted way a city has the ability to adapt to changing conditions.

To apply OA, the first step is to gain insight into the lifespan of urban assets and based on their geographical location to map areas where multiple assets reach the EOLC at a similar future moment in time (see also CHAPTER 2). In other words, an assessment of the spatiotemporal distribution of urban renewal cycles allows to plan and integrate future climate change adaptation measures in redevelopment projects. In this chapter, a mapping methodology is presented and applied to a flood prone region in Bangkok, Thailand. This chapter describes the methodology, the case study area where this methodology has been applied, the outcomes and its potentials for wider application in practice.

Along with the methodology, a model is introduced which to compare the various spatiotemporal characteristics of adaptation opportunities: EOLCs of single adaptation interventions to clustering interventions involving large scale redevelopments of neighborhoods have been considered.

3.2 Lifecycle assessment methodology

All over history, cities have been adapting to new conditions by integrating physical as well as socioeconomic and cultural conventions into 'urban renewal'. Hygiene, social rehabilitation combined with mass production for instance drove much of the urban renewal during the early industrial age. Earlier, massive fires drove a transition to a compartmentalization of blocks and neighborhoods as well as a transition to brick constructions instead of wood. Brown and Wong (2009) identified urban transitions as an evolutionary process in urban water services. The potential of explicitly integrating climate change adaptation into urban renewal cycles was emphasized by Zevenbergen (2008 and 2010). However, an operational method for this approach is still lacking. This also implies that the identified potentials of OA in the literature remain merely theoretical.

To estimate the future EOLC of urban assets, the first step is to acquire knowledge about the construction period of those assets. Typically, such information is available in a GIS-database or needs to be obtained through an examination of the cadastral records, a survey or if no information is available, through an educated guess by assessing the age of comparable neighboring structures. The second step consists of estimating the actual lifespan of the assets, i.e. the length of the useful life of a depreciable fixed asset. Here information can be obtained from specifications, guidelines, literature reviews as well as from expert opinions. A complicating factor is that regional differences are substantial, especially when considering assets at the building scale. For instance, the lifespan of buildings (houses and apartments) in northwest Europe typically exceeds 100 years, while in Japan these buildings are generally depreciated in only a few decades (Ronald, 2008). Furthermore, the lifespan differs significantly between different types of assets (buildings, infrastructure, public spaces) or even between assets of the same class.

By combining the construction period with the estimated lifespan, an estimation can be made of the EOLC. When applied to a larger region, the cumulative EOLCs at each future point in time can be regarded as the potential AR of the area: the annual fraction of all assets in a given region that reaches the EOLC at a given future point in time. Depending on the spatiotemporal distribution of the assets in combination with their estimated lifespan, Adaptation Rates may differ significantly. Fig 3-1 illustrates the AR for three distinct theoretical cases referred to as 'autonomous adaptation', 'clustered adaptation' and 'top-down adaptation' (or formal adaptation), resp.

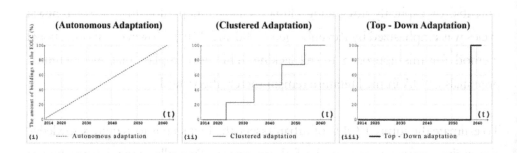

Fig 3-1. Progression of assets estimated EOLC for 3 different base cases: autonomous (left), clustered (center) and top-down (right)

Autonomous adaptation (Fig 3-1, left) involves the gradual adaptation (to meet for instance a new protection standard or flood level) through retrofitting or replacement of individual assets. In this theoretical situation every year a similar number of assets reach EOLC. Clustered adaptation. (Fig 3-1, middle) comprises adaptation of a cluster of assets through retrofitting or replacement of a group of assets that are located in close proximity as they will collectively reach EOLC at the same time. Clustered adaptation may provide opportunities to scale-up adaptation. This will allow to adopt a more integrated, design-based approach. This in turn may lead to a potential reduction of adaptation costs and further unlocking of additional benefits (e.g., Ashley et al., 2007; Jha et al., 2012; Naumann et al., 2011; Rozos & Makropoulos, 2013). Finally, top-down adaptation concerns an adaptation scheme which is integrated into the planning and design of a complete neighborhood or region. These interventions are part of a city or regional adaptation strategy and are therefore also referred to as 'planned adaptation' and are typically found at the fringes of the city.

While option I and III are providing the theoretical boundaries for OA, clustered adaptation might best reflect the actual conditions on the ground where typically a few houses or a block of houses, the adjacent sidewalk and street as well as the underground infrastructure are refurbished, upgraded or replaced.

The opportunities for clustered adaptation are determined by the number of assets with coinciding EOLCs in a fixed region. Yet, when strictly applying estimated lifespan to assets, their individual depreciation schemes and subsequent EOLCs might differ marginally, e.g., the EOLC of a neighboring building might be 2 years after that of the building in question while the facing street is already depreciated 3 years prior. This expected time gap in EOLs (EOLC-Gap) can be overcome by introducing some margin, thus synchronizing the EOLC of adjacent assets within a certain threshold.

To automate this assessment, a GIS-based model has been developed based on a commonly used clustering method. This method applies a fixed-radius nearest neighbor search (Bentley, Stanat, & Jr, 1977) to cluster assets based on similar EOLCs and to calculate the mean EOLC of that cluster. The method is conceptualized by using two parameters:

- Maximum clustering radius r. The maximum search radius from a cluster center for which new candidate members (i.e. new assets are sought). This parameter represents the spatial component of the cluster;

- EOLC-Gap g: maximum difference in time between the EOLC of the mean cluster age and a given asset within d. This parameter represents the temporal component of the cluster.

More formally, membership of a cluster **C** located at *(x,y)* and with an EOLC *t* for a candidate asset **A**, is provided when the difference in EOLC does not exceed the EOLC-Gap **g** and the Euclidean distance does not exceed the maximum clustering radius *r*.

$$\left.\begin{array}{l} |A_t - C_t| \leq g \\ \sqrt{(A_x - C_x)^2 + (A_y - C_y)^2} \leq r \end{array}\right\} \quad (1)$$

Application of this method illustrated in Fig 3-2, which shows an example of a neighborhood populated by 16 buildings with different EOLCs. Cluster **C** includes 3 buildings within the maximum clustering radius *r* which do not exceed the EOLC-GAP *g*.

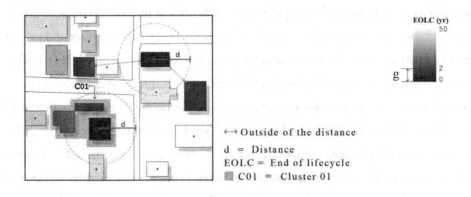

Fig 3-2. Clustering system diagram

Obviously, the values for the maximum clustering radius and EOLC-Gap influence the resulted cluster sizes. The actual choice of these threshold values depends primarily on the asset density distribution. In order to create spatially homogeneous clusters, the maximum cluster distance in a high-density urban area should be significantly smaller than in low density suburban area. These issues will further be elaborated in the upcoming sections, where the method is applied to an actual case study location. The method is implemented into a model developed in Grasshopper and Rhino-3D (McNeel, 2010; Camporeale, 2013; Tedeschi, 2011; Wagdy et al., 2015), but can be easily transferred to other platforms.

3.3 Case study: Lad Krabang

Over the course of history, Thailand is increasingly affected by floods (Manuta et al., 2006; Rattanakanlaya et al., 2016). Today, floods still occur primarily during the monsoon resulting in both fluvial as well as urban floods. In Thailand's cities, the provision of a well-functioning drainage system is extensively impaired by unplanned urban growth, which also limits efficient conveyance into canals. The latter are often encroached due to land grabbing and blocked due to poor solid waste management (Nakayama et al., 2013).

Bangkok is a megacity with a population of 9.6 million in 2018 (Hara et al. 2005). Extensive population growth has resulted in a reduction of the number and size of the areas that serve as flood retention (e.g., agricultural farm land). Uncontrolled urban growth has changed the function of residential and retail areas. With an ever-decreasing infiltration and storage capacity, flooding becomes more frequent and severe (Yokohari et al., 2008). In the aftermath of the extreme flood event of 2011, the government provided financial compensation in the form of subsidies (Johannessen et al., 2014). These subsidies were primarily destined to repair and refurbish buildings that were damaged during the event, some of the buildings were inundated for over 2 months (Poapongsakorn et al., 2013). Furthermore, the Department of Drainage and Sewerage of the Bangkok Metropolitan Administration (BMA) has developed a master plan to reduce future flood risk by constructing large scale drainage infrastructure works, including a 200 km underground 'super drain' that acts as a bypass to relieve Bangkok from peak discharges developed further upstream. Completion of this project will likely take 20 years (Bretz, 1998; Dhakal & Shrestha, 2016). Yet, at a more local level, drainage capacity is decreasing rapidly as is the case in the district of Lad Krabang, which serves as the case study area in this chapter. This district is representative for many of the areas located in the fringes of the Bangkok metropolitan zone.

Lad Krabang is a former agricultural area that has initially been designated to serve as peak storage where locally developed runoff is stored (Keokhumcheng and Tingsanchali, 2012). Due to its proximity to an international airport (Suvarnabhumi) as well as to a vast logistics zone, the area has become increasingly attractive and is still growing significantly (Nasongkhla & Sintusingha, 2011). Although since 1992 a major portion of the area is excluded from development, actual enforcement of planning regulations has been poor. While, in the late 70s about 60% of Lad Krabang consisted of agricultural farm land and flood storage, this number dropped to about 27% in 2016 (see in Fig 3-3(ii)).

The majority of the constructions has been made from concrete although a significant portion (about 20%) consists of wood. Additionally, Lad Krabang comprises a series of sub-districts which have been built in the period 1992-2012.

Flooding due to local rainfall in Lad Krabang occurs annually during the wet season. Fig 3-3(A, B, and C) depicts the flood level that occurred in October 2011. Shortly after the peak of this rainstorm, inundation levels reached up to 50 or 60 centimeters causing damage to business as well as traffic interruption. It took around 2-3 hours for the area to become flooded. In recent years, an increasing number of homeowners has adapted their houses by elevating ground floors in anticipation of future flood events. The height of the ground floor has been recently upgraded from around 0.80 to 1.20 meters based on the experience gained during the previous extreme flood of 2011 (Dutta, 2011; Guha-Sapir et al., 2012). Unfortunately, these autonomous driven adaptation measures will likely not be sufficient to prevent future flooding caused by extreme rainfall (Thomalla et al., 2017). Moreover, there is a risk that these measures are maladaptive: they could aggravate the risk of flooding of neighboring buildings and infrastructure (Limthongsakul, 2017; Escarameia et al., 2016).

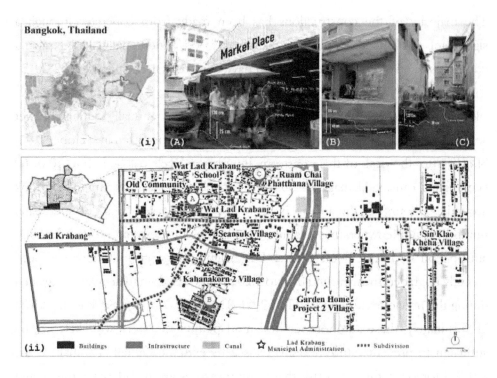

Fig 3-3. (i) Urban plan of Bangkok, Thailand, (ii) Case study area (1,859 buildings), (A) Local market, (B), and (C) Housing areas

An inventory of the building stock in Lad Krabang has been conducted based on information provided by the Lad Krabang District Office in 2014 covering the construction period from 1980 to 2012. The annual construction volume for this period is shown in Fig 3-4. The distribution has a mean of 56.3 building per year. While a substantial variation is observed, no significant trend-wise changes can be identified. While it is likely to assume that the growth of the area is driven by the development of the international airport, this is not reflected in the number of buildings constructed in the years before the airport has been constructed. Furthermore, since the exact size (number) of the building stock has not been monitored over the years, it is unclear which percentage of the constructions is due to the replacement of older buildings. This data only exists for 2014, when the Lad Krabang building stock consisted of 1,859 buildings.

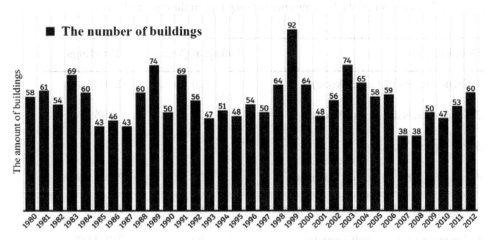

Fig 3-4. Annual construction volume 1980 to 2012

The 2014 building stock-data is categorized by building types and primary construction materials. Using information offered by the Agency for Real Estate Affairs, which has published data obtained from the Bangkok region (Pornchokchai, 2015), the average lifespan was determined for each building type.

Table 3-1 shows the building types and their corresponding lifespan used in this study. The lifespan of 62% of the total number of houses (the combined types 5, 6, 7 and 9) is estimated at 50 years. Detached wooden houses on the other hand (types 1, 2 and 3) have a 20 years lifespan corresponding to 22% of the total building stock. The remaining 16% consists of mixed-material houses with an estimated lifespan of 25 years. Finally, a marginal portion consisted of public housing blocks and warehouses with lifespans of 45 and 30 years, respectively.

Table 3-1. The number of the buildings per building type in the case study area.

Building Types	Lifespan (Year)	Buildings
1. Single detached 1 Floor (Wood)	20	250 (13.4%)
2. Single detached 2 Floor (Wood)	20	97(5.2%)
3. Pile House (Wood)	20	63 (3.3%)
4. Housing (50% Wood and Concrete)	25	289 (15.5%)
5. Single detached - Townhouse (Concrete)	50	877 (47.1%)
6. Apartment 1-5 Floors (Concrete)	50	242 (13.0%)
7. Social Housing and Flat	50	24 (1.2%)
8. Public Building (Low-rise)	45	6 (0.03%)
9. Public Building (High-rise)	50	4 (0.02%)
10. Warehouse	30	7(0.03%)
Total (Buildings)		**1,859**

Source: Pornchokchai (2015)

3.4 Assessment of the adaptation potential using the clustered adaptation approach

The AP has been assessed using the clustering method described in Section 3.2. The clustering method was performed using an EOLC-Gap of 5 years and a clustering radius of 100 meters. Due to the scale of the urban infrastructure, the average age of the buildings was used in the process. The outcomes are presented in Fig 3-5, in which the potential AR and the adaptive potential (cumulative AR) are shown for buildings that are captured inside clusters, or outside clusters (autonomous adaptation). The figure covers a period of almost 50 years, which ensures that all buildings reach the EOLC by the projected horizon of 2064.

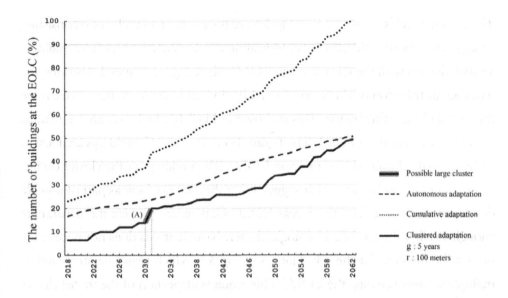

Fig 3-5. Adaptation Potential (cumulative AR) and AR of Lad Krabang District for the period 2018 till 2062

It follows from this figure that the number of buildings which are captured in the clusters is lower than the number outside the clusters. So, especially in the near future the majority of depreciated buildings are better suited for autonomous (i.e. stand-alone) adaptation. Yet after 2048, the AR for clustered buildings increases resulting in an almost equal share beyond 2060, which marks the upper bound of the adaptation horizon (918 against 942 buildings respectively). The sudden jumps in the black (lowest) line represent clustered adaptation of a relatively large number of buildings. In other words, these jumps indicate a large number of buildings reaching their EOLC simultaneously, and thus can be potentially adapted in one strike. In Fig 3-5, this is illustrated for the interval 2030-2031 (segment - A) where 117 buildings reach the EOLC. Although less dramatic, the line representing the clustered buildings shows many jumps, suggesting significant cluster sizes. As expected for a relatively differentiated neighborhood with a relatively even annual construction volume (see Fig 3-4), the AR for buildings not captured in a cluster is relatively stable.

The question still is, if the jumps in Fig 3-5 are the result of a small group consisting of large clusters or if the jumps are indicating the combined effect from multiple smaller clusters with the same average EOLC. Since Fig 3-5 is merely showing the temporal distribution of cluster-associated EOLCs, it is also important to examine the spatial cluster distribution. This is shown in Fig 3-6, which indicates individual EOLC-based clusters for 2031. The figure shows that all 117 buildings are located within a single cluster with a common EOLC. The buildings in the cluster do not show a strict urban layout, which suggests that they are not built as part of a single development project. Yet, the 5 year EOLC-Gap seems to relax the constraints enough to align the EOLCs in a single cluster. Note that this does not necessarily need to be the case for the other jumps in Fig 3-5, which might be attributed to multiple clusters reaching the EOLC. This requires inspection of the spatial cluster distribution for each year in the projection period. Furthermore, it shows the importance of focusing on the spatiotemporal dimension instead merely addressing the Adaptation Rates (temporal) or spatial cluster distribution for a static point in time.

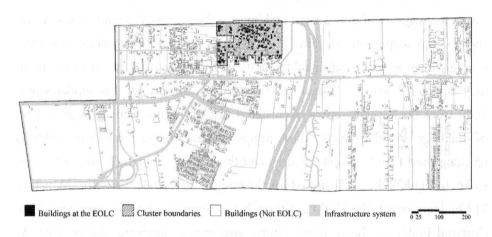

■ Buildings at the EOLC　▨ Cluster boundaries　☐ Buildings (Not EOLC)　▨ Infrastructure system　0 25　100　200

Fig 3-6. Spatial cluster distribution in 2031

Apart from the spatiotemporal distribution of the building's EOLCs in Lad Krabang, the observed cluster sizes depend on the parameter values used for the EOLC-Gap and the clustering radius. To assess the influence of increasing the values of these key-parameters changes to the clustering behavior of the algorithm, three additional model runs have been performed with the EOLC-Gap set at 10 years, a clustering radius set at 200 meters as well as a run with both these settings respectively. The outcomes are shown in Fig 3-7 and set against outcome of the initial parameter setting shown in Fig 3-5.

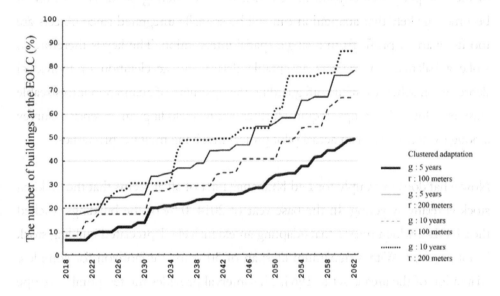

Fig 3-7. Adaptation Rate using 4 different parameter settings

It follows from Fig 3-7 that increasing the parameter values, the algorithm becomes more "greedy", i.e. relaxing the temporal and spatial constrains for cluster membership results in larger clusters and an overall increase of the fraction of buildings captured in clusters. The increased cluster radius (200 meters) results in a larger portion of buildings captured in clusters, indicated by the substantially larger jumps compared to the reference value. A similar behavior can be observed when setting the EOLC-Gap to 10years. In conclusion, the effect of the parameter changes is distinct and leads to the formation of different clusters. This can be observed when the two (dash-dot line) are being combined, which result in extensive jumps indicating a cumulative effect of the relaxed constraints.

The parameter values should reflect operable conditions for scaling up adaptation in Lad Krabang. Obviously, larger clusters can be identified when increasing the EOLC-Gap further, possibly in combination with a larger search radius. Yet, a large EOLC-Gap in practice means that many of the building owners have to either initiate adaptation well before their properties are depreciated or extend the lifespan of their property well beyond the EOLC. If the clustering radius is too large, it becomes unlikely that adaptation can still be spatially integrated since objects are too far apart to profit from a single spatial intervention. The key is therefore to strike a balance that reflects acceptable delays and acceleration of individual depreciation schemes, while safeguarding the upscaling of adaptation into sizeable clusters. Optimal settings require further study including an extensive survey among owners or even an actual pilot project involving multiple properties.

Now what does this imply for Lad Krabang? First of all, it seems that the building stock is relatively young. In the base year of 2014, 0 % of the buildings reached their EOLC, which means that adapting an extensively depreciated building stock is not required. With a maximum lifespan of 50 years, the horizon for complete adaptation of the area is set to 1964. This interval provides the temporal envelope in which OA can take place. Furthermore, the upper boundary is determined by autonomous driven adaptation of individual assets, with its earlier maladaptation and redistribution of risks as possible collateral effect. The lower boundary is determined by a top-down driven adaptation approach which implies a long period of inaction in which flood impacts are suffered by the Lad Krabang community. All buildings reached their EOLC in 2064, indicating that a large-scale adaptation can be implemented in a possibly transformative and integrated manner. In Lad Krabang, OA could start immediately on a small fraction of buildings which already reached the EOLC. However, as of 2021 larger cluster reach the EOLC, which provide opportunities for scaling up adaptation. This trend continues till about

2031, from which on cluster-based adaptation opportunities appear to stagnate until 2048, from which the speed seems to pick up again.

The clustered adaptation strategies might imply the best of both worlds: they do not require a very long period of inaction, but also integrate adaptation measures into sizable development involving multiple buildings, plots, streets and adjoining drainage infrastructure. Depending on the locations and the characteristics of those areas, a wider portfolio of integrated measures might be available and an opportunity for more cost efficiency as well as integration of livability and other benefits outside the scope of flood protection (i.e. multiple benefits) could be attainable. Obviously, a more in-depth analysis is required to evaluate an actual portfolio of measures as well as the possibilities for combining the measures with additional functions in order to maximize the added value to the neighborhood.

3.5 Discussion and conclusions

This chapter proposes to start thinking in a more systematic and integrative manner about urban regeneration and climate change adaptation. Both have become an acute challenge for most cities nowadays. While it is straightforward to propose climate proofing measures for new urban areas, adapting the existing city requires new approaches, such as OA, to transform the urban fabric in order to keep up with the future challenges such as imposed by climate change.

OA is a climate adaptation approach, which is based upon an understanding of the AP and in which adaptation measures are integrated into urban renewal cycles. The AP depends on (i) the lifespan and lifecycle of the individual elements of the existing urban infrastructure and (ii) the potential synergies which may arise from taking these individual elements together in one aggregated and integrated adaptation intervention (cluster).

By adopting an asset management planning approach, a rationalization of asset depreciation is applied where adaptation measures are implemented when assets reach the end of their functional life; their EOLC. Especially in cities, this could lead to the implementation of more gradual and less intrusive adaptation responses as well as to a potential cost reduction compared to formal, stand-alone adaptation measures. Furthermore, this approach is inherently more flexible than traditional methods (related to both the pace of implementation and the type of measures selected) thus enabling a more resilient response to future climate uncertainties. However, OA might limit the opportunities for large scale adaptation projects. Depending on the differentiation in the expected EOLC of assets, adaptation might become fragmented and will be limited to projects involving individual assets. Alternatively, the proposed clustered adaptation may not only lead to better integration, the resulting synergies can also contribute to a further co-benefit while expanding the portfolio of possible adaptation measures. For instance, if the interventions are limited to flood proofing of individual buildings, this might over time be a relative expensive option compared to an integrative flood proofing intervention of a complete group of buildings, adjacent streets and public space. In order to test this concept, it is essential to gain insight into the spatiotemporal distribution of the EOLCs of urban assets in actual conditions. The estimated EOLCs of assets located in specific regions can be synchronized by allowing a margin, the EOLC-Gap. This means that some assets are depreciated prior to their expected EOLC while for others, their lifespan is extended.

This chapter provided a first step to an operational method of OA by focusing on the life cycle aspects and applying a GIS-based clustering method on a district of Bangkok, Thailand which is frequently suffering from floods. The clustering method is driven by two main parameters expressing the spatiotemporal aspects of OA. Moving between a completely autonomous driven approach in which individual buildings are adapted one by one and a top-down approach in which a complete district is redeveloped all together, a series of clustered adaptation opportunities were identified.

With life spans ranging between 20 and 50 years, it is expected that the relatively young building stock will be completely depreciated around 2060. Within this time horizon, major adaptation opportunities appear in the 2030-2040 interval and in the 2050-2060 interval. This would allow for the development and implementation of a number of tailored adaptation plans that include sizable zones consisting of large sets of buildings. To explore the scope of possible measures, more detailed studies are required that focus on the spatial characteristics of the clusters, the expected climate change impacts as well as many social economic and technical aspects of the buildings in question.

The implementation of a cluster-based integrated adaptation approach requires a number of enabling factors that might not be readily available in Lad Krabang. In practice, mainstreaming adaptation opportunities are dictated by the depreciation of high-level assets such as spatial and temporal damage-cluster identification, economic value versus expected annual damages, and aggregate damage values for individual neighborhoods (Karimpour et al., 2014; Lotteau et al., 2015; Veerbeek & Zevenbergen, 2009). Apart from a rationalized process of asset management, the process requires a design-driven approach facilitated by architects or planners with a strong presence in the community. In turn, building owners are required to organize themselves within the identified clusters in order to negotiate and evaluate proposals. This is not straightforward, since current ownership structures are fragmented. Even then, there might be many obstacles that hamper a successful implementation. Further exploration of these aspects is outside the scope of this chapter but will be further developed in a forthcoming article.

One of the major assumptions in this study is that with reasonable EOLC-Gaps and search radii, EOLC-driven clusters can be identified. For the Lad Krabang area future potential adaptation projects have been identified that could be part of an integrated, cost efficient and multi-benefit generating adaptation approach for this area. Further detailed study will be required to assess the OA potentials for other areas in the Bangkok metropolitan area with different spatiotemporal characteristics. In compact urban areas developed in a relatively short period of time, distinct adaptation opportunities may appear in which a significant fraction of the area can be adapted at the same moment in time. Yet, in other areas characterized by a dispersed building stock and constructed gradually over time, such opportunities may be absent. It is therefore recommended to apply this methodology to other areas of Bangkok. These insights could be used to further assess the wider potentials of OA for this city and beyond.

CHAPTER 4
Developing adaptation pathways
Case study Kehanakorn

Operational method
CHAPTER 2

Assessment and analysis
CHAPTER 3   CHAPTER 4

Assess adaptation potential	Identify adaptation options		Assess current and future flood risk
Lifecycle assessment methodology	Catalogue and classification	↔	Flood modelling/empirical data (beyond scope of this study)

Developing adaptation pathways
CHAPTER 3 + **CHAPTER 4**   CHAPTER 3 + **CHAPTER 4**   **CHAPTER 4**

Define clusters		Map opportunities		Identify pathways
Urban density, synergies/co-benefits	↔	Selection of options	↔	Decision Tree method

Towards a strategy
CHAPTER 5   CHAPTER 6

Evaluating flexibility	Governance
FIA method (measuring flexibility)	Interviews and survey, Ex-ante evaluation

4.1 Introduction

Cities are dynamic systems. Understanding and planning the urban system are key to applying flood adaptation measures in an urban area. This chapter, the concept of adaptation pathways, a sequence of adaptation measures to be implemented over time, has been introduced. In this study adaptation pathways are considered as a key component of an urban planning strategy to guide sustainable (climate proofing) regeneration of cities: it allows designers, urban planners and decision makers to consider a raft of adaptation measures to be implemented on the short-, medium- and long-term. It introduces and supports strategic thinking in the design process by taking into account the life cycle of buildings (Adaptation Potential), their spatial distribution (mapping) and potential synergies and co-benefits to combine and integrate interventions (clustering). In the current design practice, urban designs are considered as snapshots in time. These static plans are produced to depict and explore the shape, organization and function of an object or system (such as a building or urban landscape) before it is realized. The notion that urban objects or systems change over time and that adaptation measures have a "sell-by date" requires to consider a reframing of the classic, static designing approach. Single designs should be considered as discrete steps of an ever evolving (adaptive) process and should be based on the acknowledgement that conditions may change over time in an unpredictable way. Hence, a more dynamic, iterative design approach will be required which explores and depicts how the object or system may evolve over time and what the consequences of a design decision will be for future adaptation options, such as creating an infrastructural lock-in which constrains the number of future adaptation options. In this study adaptation pathways serve to guide this dynamic design approach.

Bangkok is suffering from both frequent low impact events (rainfall) as well as from rare (frequencies lower than 1:50) extreme flood events (combination of fluvial and pluvial). In this research the focus is on the first category as there is much to gain using local interventions (from landscape to building level) to mitigate their impacts as opposed to mitigating the risk of extreme events which also require large scale interventions (monkey cheeks, embankments, barriers etc.). This also means that for an assessment of the AP of this category of events, the use of 'rough' assumptions (based on historical data and flood modelling) about maximum flood levels would generate meaningful and sufficiently robust outcomes.

This chapter describes the method and its underlying thinking of developing adaptation pathways using the Decision Tree Method and describes a case study in which the method has been applied following a dynamic design approach. A decision-tree is conceived as a single diagram that represents all possible flood adaptation measures at each spatial scale level, ranging from individual buildings, to clusters of buildings, and finally to the whole area. the Decision Tree Method uses a decision tree algorithm (Grajski, Breiman et al., 1986; Quinlan, 1996; Kraslawski, 2013) for LCA to identify flood adaptation measures in a defined area. Hence, the aim of this chapter is to demonstrate this novel approach to support dynamic design for strategic adaptation and planning by presenting designs capturing a sequence of measures to be implemented at different spatial scales over different time intervals in the case study area.

The adaptation strategy process is depicted in Fig 4-1 and combines the required methods and builds upon the lifecycle assessment from CHAPTER 3. The adaptation strategy process involves three components: i) the input data, ii) the methods and principles which are processing these data, and iii) the resulting outputs. The type of information required for to the input data comprise of (i) area characteristics (such as flood features[1], building type and age and size of the area) and (ii) potential flood adaptation options (the catalog of measures). The methods and principles component involve (i) opportunity mapping, (ii) design approach, (iii) design criteria, and (iv) Decision Tree method. These are explained in this chapter and the next. The outputs of the adaptation strategy process are (i) adaptation pathways, and the resulting (ii) adaptation strategy. The adaptation pathways provide the pillars of the adaptation strategy. A key feature of the strategy is a dynamic design which involves a sequence of 'snapshots in time' depicting the transformation process of a cluster over time. Important to note here is that in the process step of strategy development the features and adaptation pathways of neighboring clusters, which are physically connected, have to be considered in this process as well. The ultimate aims to deliver an integrated, dynamic design which spans several clusters to maximize synergies.

---

[1] Over the past 2 decades, after the big flood in Thailand of 2011, numerous flood risk studies have been conducted. These studies vary from river basin level (fluvial/pluvial flooding) to urban areas (stormwater flooding). In this research existing flood risk maps of Bangkok and observations (max. flood depths) by local stakeholders and authorities have been used (sources and data derived from Royal Thai Survey Department and Thailand's Geo-Informatics and Space Technology Development Agency (GISTDA)).

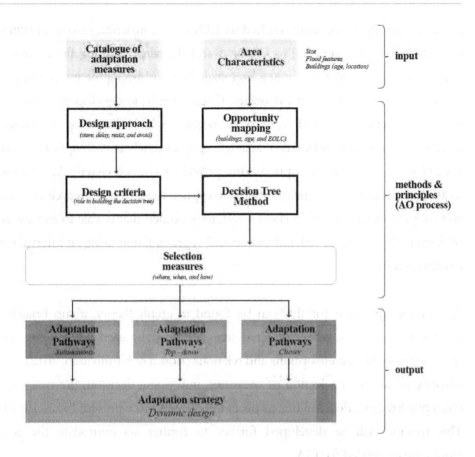

Fig 4-1. The adaptation strategy process underlying the OA approach

4.2 Decision Tree method

4.2.1 Sequential representation of measures

OA implies the identification of adaptation opportunities at the EOLC, i.e. the availability of a suitable set of adaptation options for a given asset at a future point in time when an asset is depreciated. One of the objectives of OA is to provide insight in the adaptation opportunities over a longer period of time. Consequently, a long term assessment needs to cover multiple lifecycles and subsequent sets of adaptation options. A constraint is that the choice of future adaptation options might be limited by the selection of options at an earlier interval, i.e. at a previous point in time where the asset reached its EOLC. For instance, once a building is dry-proofed (i.e. the building's exterior is sealed against entering flood water) it makes no sense to wet-proof it at a later stage by refurbishing the interior in order inundation does not cause any damages. Consequently, to develop an operational method a representational system is needed which provides an ordered representation of sequential sets of adaptation options, where each option provides a seed for a new sequence of options, thus providing a nested system. An additional requirement is the representation of time to express the different lifecycles of assets and adopted measures. So, in short a system is needed that is able to express sets of discrete choices in an ordered (sequential) representation using an interval scale to express time.

A common approach for this can be found in graph theory, a sub branch of discrete mathematics, where directed trees are typically used to express search spaces describing discrete options and relations (Rosen & Krithivasan, 2012). Since adaptation requires a selection of measures, implying a decision, the tree can be conceptualized as a decision tree using an interval scale to represent future EOLCs. This method will be developed further to further accommodate the actual requirements needed for OA.

4.2.2 Definition and features of a decision tree

A common definition of a **decision tree** is "a decision support tool that uses a tree-like model of decisions and their possible consequences" (Quinlan, 1987). It is used as a visual and analytical decision support tool to display a structured sequence of options. A decision tree typically consists of nodes (decision nodes, probabilities and end nodes) and edges:

- *Nodes*: points at which a measure or a number of measures can be chosen and applied;
- *Edges*: each directed edge of the graph represents a choice for a specific flood adaptation measure. The length of this edge represents the lifespan of a

single measure or asset which the time required for that particular measure to be applied.

Subsequently, one can identify branches: any subtree except the tree itself represents a branch of the original tree;

- *Branching factor*: the number of edges that originate at a node, expressing the number of possible options; the branching factor of a node represents the number of child nodes it has. A child node, is a node of the graph that is connected by a directed edge to an initial node, which we will call the parent node. For example in Fig 4-2, node N0 has a branching factor of 1, because it is connected to a single other node, while node N1 has a branching factor of 3, having 3 children as connected nodes. Nodes N5, N6, and N7 all have a branching factor of 0. These are the so called "leaf nodes", as they basically define the 'extremities' of the tree.

- *Distance*: The distance of a node to one of its children is simply the length of the edge connecting the parent to the child (lifespan of the option), which in this case signifies the number of years. In our Fig 4-2, this is marked as the red line labeled "Distance" between nodes N1 and N4. It is worth noting that in our particular example, the distance between N1 and N4 is the same as the distance between N1 and N2 or N1 and N3.

- *Depth*: distance between two nodes expressing the total period that is covered between the points in time when measures can be chosen; the depth of the node is defined as the maximum length of a path starting from this initial node and ending with any leaf node of the tree. A path is basically an enumeration of graph edges, where each two consecutive edges share a node. Conceptually this represents a possible option for applying different flood adaptation measures successively, starting from an initial moment in time. For example in Fig 4-2, node N0 has depth N which is defined as the sum of the edges (N0, N1), (N1, N2) and (N2, N5).

Having a schematic representation of the decision-making process provides a good oversight of the future possibilities of adaptation. As mentioned before, even when this tree grows and becomes sufficiently complex to be unreadable by humans, we can deploy an automated process to explore and identify pathways and measures. For this, we need to describe the characteristics of the decision tree, expressed as a formula, as it was done for example Appendix C.

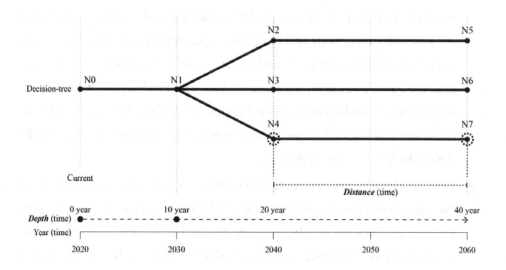

Fig 4-2. The concept and formula of the decision-tree

The construction of a decision-tree consists of two main steps. Firstly - sequencing adaptation opportunities as a function of time and secondly - creating the decision-tree. The decision-tree is constructed based on selecting the appropriate flood adaptation measures. In the next step, the possible sequence of measures is being determined. At each EOLC, this decision-tree has different branches that represent the available choices, according to the specific lifespan of each individual measure. This results in a specific strategy corresponding to the area where the adaptation opportunities exist.

The first step, mapping adaptation opportunities, aims to define the relationship between an area and the potential adaptation measures, which is captured by an algorithm. The process of mapping adaptation opportunities as a function of time is described in CHAPTER 3.

4.2.3 Selection adaptation measures

The flood adaptation measures catalog, which is used in this study, is based upon inventories and catalogs of adaptation measures developed by Jäger et al., 2015; Meyer, 2018 and Zandvoort et al., 2019. A recent example of such a tool which is successfully used by urban designers and engineers is the *climate adaptation app* (see www.climateapp.org).

In this study four principles have been adopted to prioritize and select adaptation measures of the catalog. These principles are widely used as guiding principles in water sensitive design, water resource management and FRM (such as Stumpe and Tielrooij, 2000). The four principles are: store, delay, resist, and avoid (Fig 4-3):

- Store (**S**): The Store type of intervention/measure has the function of retaining storm water. Additionally, measures to store or retain water in periods of rainfall could also serve additional functions and be integrated into the design resulting in multi-functional playgrounds and wetlands. Storage/retention basins are designed to store excess runoff water from urban drainage systems during wet periods, primarily when the runoff exceeds the discharge capacity of the urban drainage system. A settling tank is often an additional component to the measure to prevent polluted runoff from being discharged into the surface water. For example, natural wetlands function as water retention basins, sediment traps, and wastewater treatment areas by filtering and immobilizing harmful microorganisms. Wetlands can be implemented with or without these additions depending on the needs to improve the treatment capacity. As extra treatment capacity can be added to those systems, they have a certain degree of flexibility.

- Delay (**D**): The Delay type of intervention is represented by measures in the domain of green-blue infrastructure, and it is used to absorb and delay the stormwater from entering the pipe drainage system, thus limiting the volume of discharge. A typical example of an intervention of this category is adding trees to the streetscape. They slow down the speed at which the water moves down horizontally across the landscape surface. Trees provide shade and reduce heat accumulation through evapotranspiration. At the scale of a single building, a green or living roof can be partially or completely cover rooftops with vegetation. A water square can also be designed as a delay function; the water is temporarily stored and is slowly discharged into the existing pipe drainage system.

- Resist (**R**): The Resist type of intervention can be divided into two categories depending on the scale level of deployment: architecture or landscape level. At the landscape level, the intention is to prevent water entering a building. At the architecture level a wet or dry proofing approach can be followed, which either completely allows or respectively completely prevents water entering the building.

- Avoid (**A**): The Avoid type of adaptation measure involves an approach that elevates critical elements of a building or infrastructure above potential flood levels. They can be positioned on piles or on raised ground, such as traditional dwelling mounds. Alternatively, the ground floor can be designed to fulfill a secondary function such as a storage space or a garage, using the water-resistant construction methods described above. This way combined land use is possible, making room both for water storage and urban development. Furthermore, some public utilities or vital infrastructure elements could be located in vulnerable flood-prone locations, in these cases relocation to higher ground is an option to reduce flood risk.

The landscape and architectural level need to be taken into account in an integrated way (systems approach) when developing a strategy which aims to increase flood resilience (Zevenbergen et al., 2008). While the landscape level is focused on large scale (neighborhood, district) interventions such as embankments and green corridors, the architecture level is focused on individual blocks and buildings. Designing a solution that works on a large scale (landscape) discloses more degrees of design freedom than on a small scale (architectural). That is because large scale interventions may benefit from synergies (economy of scale, creating additional multiple functions) when interventions at architecture level are combined and lifted to the landscape level. Therefore, in this study, an additional design principle "intervene at the highest scale where possible" has been followed.

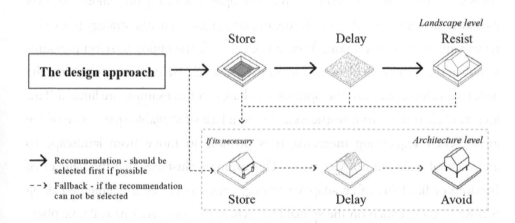

Fig 4-3. The design approach (water sensitive design)

4.2.4 Setting the decision-tree rules

The approach for creating the decision tree allows to define flood adaptation measures in a time sequence (Adaptation Pathway approach). Fig 4-3 shows a decision tree where the presented flood adaptation components are ranked on the basis of the design principles mentioned above. The arrows in the figure show the order in which the individual components are ordered in a sequence, revealing the available adaptation options.

In addition to the principles for prioritizing and selecting potential measures, the lifespan of assets also plays a role in building the decision tree. Hence, another rule should be added which involves the lifetime of assets (see also CHAPTER 3). Fig 4-4 on the left (a) presents the application of this rule: every time when an asset reaches the end of its life-cycle an intervention will take place, which could be upgrading of replacement. This can be performed either at the level of the landscape or at the level of the architecture. The figure presents the available measures as a function of time to be implemented, in this example with equal weight, either at the architectural (marked with an 'A') or at the landscape (marked with an 'L') level. Fig 4-4 on the right (b) shows when the principle of 'intervening at the highest spatial scale' is followed: flood adaptation measures are 'preferentially' applied at the landscape level. In this example, following the landscape level pathway, there are no lock-ins as future opportunities to change strategy (measure) remain after each intervention. At each node there is the option to select measures at either the architectural or at the landscape level. By contrast, following the architecture level pathway, the options for change in this example are limited. This lock-in likely results from restrictions due to a lack of available space (size of the area). When sequencing measures, it is possible to move from landscape to architectural scale but not vice versa. This example illustrates that the degrees of freedom or flexibility of an adaptive strategy (measures to be implemented along pathways) is depended on the spatial level where the interventions will take place: the higher the spatial level, the higher the number of measures to intervene and the lower the risk of creating lock-ins.

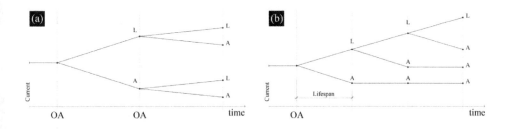

Fig 4-4. Developing a decision tree

The ordering principles (store, delay and resist, seizing the opportunity and intervene at the highest spatial level), which have been described above have been used to formulate a set of rules (algorithm) which shape the pathways of a decision tree.

An example of a such a systematic process to develop pathways using these principles is given in Fig 4-5. The preferred pathway follows the grey line which involves intervention decisions at the landscape level including adaptation measures to 'Store' water (indicated as LS points on the pathway). If 'Store' cannot be accommodated by an adaptation measure (due to technical, legal or financial constraints), an adaptation measure which 'Delay' water is preferred (indicated as LD points on the pathway), and so on. A similar line of reasoning is followed for intervening and Landscape (L) or Architecture (A) level, whereby A is chosen when L is no longer feasible.

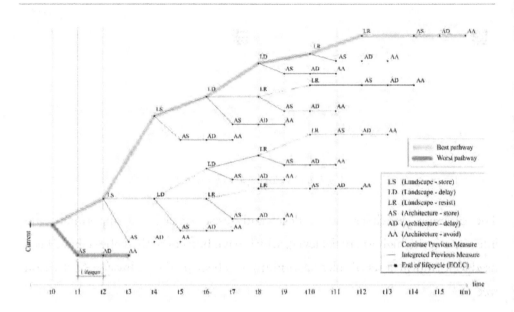

Fig 4-5. Overall process and 'best path and worst path'

Flexibility with respect to the available options is an important feature of an adaptation strategy (see also CHAPTER 2). Developing a decision tree as described above is a useful method to explore, identify and rank potential adaptation measures to be implemented over a longer time period. A sequence of potential measures, defined as adaptation pathway, can be assessed and further adjusted based on its *flexibility* (degree of risk to creating lock-ins).

The methodology of building pathways using the Decision Tree Method has been applied in the case study area of Kehanakorn Village 2, Lad Krabang district, which is presented in the following two sections (on the theory and the design, respectively). These two sections depict the different steps applied to the case study area: mapping adaptation opportunities as a function of time, defining the clusters, analyzing them, and finally integrating the result into a building program.

To develop the decision tree and strategy a computer simulation based on the design principles has been executed. For the computer modeling, the following input parameters have been considered:

1. Area requirements (square meter.)
2. Water levels (meter)
3. Flood frequency (year)
4. Lifespan of measures (year) which is based on the maintenance and operation section
5. Flood adaptation measures (catalog)

4.3 Application of the methodology

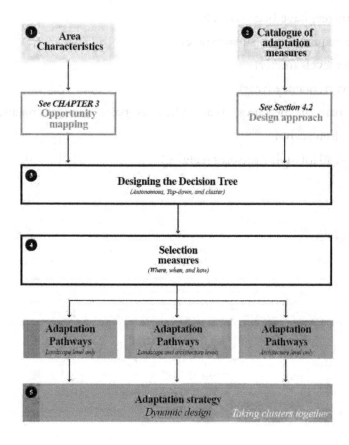

Fig 4-6. Process scheme of the methodology

In this section the methodology is applied in the case study area. Fig 4-6 depicts the overall adaptation strategy process, which consists of five-step below as already described above. The results are presented below for each step of the process scheme Fig 4-6.

1. Area characteristics

Fig 4-7 presents a detailed map (building stock) of the location of Lad Krabang district and Kehanakorn village 2 (same location as CHAPTER 3). The construction of the buildings in the Kehanakorn village 2 (192 units) started in 1990. The area has not seen any large scale government driven redevelopment projects over the past 30 years, in spite of the fact that it is located right next to Suvarnabhumi Airport. Floods are occurring every year, and their frequency is increasing. The highest flood water level recorded was set by the extreme flood event in 2011: the maximum flood water level was 0.8 meters (Keokhumcheng and Tingsanchali 2012; Singkran and Kandasamy 2016). This extreme flood event revealed that the area is highly susceptible to flooding. That is why it is used as a case study in this study.

(a) current situation of a case study area, Kehanakorn village 2

(b) Images present an area surrounding context of Lad Krabang district

Fig 4-7. Map of the study area called "Kehanakorn village 2". (a) are presented current situation and context of Lad Krbang district. Images (b) are surrounding context of Lad Krabang district

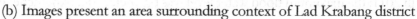

QR code (1)
Scan me – This will show the extreme flood situation at 2011

QR code (2)   QR code (3)
Scan me – These will present current flood situations 2020

QR code (1) presents an extreme flood event which has occurred 8 years ago, which affected this location. QR code (2) and (3) show the current flood situation 2020.

2. Catalog of adaptation measures

Table 4-1 presents an example of a page of the catalog of flood adaptation measures based on the design principles to be applied at both landscape and architecture level (see 4.2.2). This table shows the necessary information related to the input parameters for the simulation, notably the flood adaptation measures, area requirements, the lifespan of measures, water levels and associated flood frequency. The complete overview of the information provided by the catalog is given in https://www.climateapp.org/- (see also Appendix D).

Table 4-1. An example of (page) the catalog of flood adaptation measures.

Flood Adaptation Measures	Area Requirements (Sq. Meters)	Lifespan (Year)	Water Level (Meters)	Flood Frequency (Year)
Resist (Architecture)				
Sealable buildings (dry proof)	10	12	0.3	Every 1
Wet proofing (water resistant construction)	10	15	0.6	Every 1
Resist (Landscape)				
compartments in dike rings	1,000	20	0.3	Every 20
Elevated flood wall	500	10	0.6	Every 1
Dikes	2,000	15	2	Every 1
Floodable dike	2,000	50	1.5	Every 20
Avoid (Architecture)				
Elevated building	50	20	3	Every 1
Increase height difference between street and ground	30	20	1	Every 5
Raising the ground floor level	20	15	0.6	Every 1
Buildings (partly) Situated in the water	100	30	0.6	Every 1
Floating buildings	50	50	2	Every 5
Avoid (Landscape)				
Building on partially elevated areas	2,000	25	2	Every 1
Raising land	1,000	5	3	Every 1
Artificial islands	4,000	30	3	Every 2
Delay (Architecture)				
Sloping roof	25	10	0.3	Every 1
Green facades	10	5	0.3	Every 1
Green roof	10	5	0.6	Every 1
Delay (Landscape)				
Pocket park	850	10	0.3	Every 1
Green infrastructure	1,000	20	0.5	Every 2
Urban wetland	4,000	20	0.6	Every 1
linear park	850	10	0.3	Every 1
Park	1,000	12	0.3	Every 10
Store (Architecture)				
Airbag water storage	10	15	0.3	Every 1
Rainwater tanks	5	10	0.3	Every 1
Tank roof	20	10	0.6	Every 1
Store (Landscape)				
Water Square (plaza)	500	30	2	Every 1
By-pass creation	3,500	30	1	Every 1
retention area	4,000	20	1.5	Every 1
polder	2,000	15	0.6	Every 1

(Source: https://www.climateapp.org/- An example of (page) the catalog of flood adaptation measures)

3. Designing the Decision Tree

In CHAPTER 3, three different adaptation strategies (autonomous, top-down, and clusters) have been presented. Fig 4-8 illustrates the application of these strategies in the study area. These figures depict for each strategy the potential measures which can be implemented over time based on their EOLC. The subfigures focus on the following aspects: (a) spatial occupation, (b) temporal dimension (Decision tree), and (c) selected measures. The period considered in the Figures covers 50 years to ensure that all buildings reach the EOLC within the projected horizon of 2064.

Fig 4-8(i) shows the adaptation strategy at the smallest scale, namely at the level of a single asset – at the architectural or building level the process involves autonomous adaptation. In this example, the area covers 300 square meters and a total of five measures will be considered. The first adaptation opportunities emerge in 2027 to include two options: elevating the ground floor (avoid type) and wet proofing (resist type). In the next period, pathway number 2 creates four options, and route number 1 only two options. The diagram in Fig 4-8(ii) shows the adaptation strategy at the largest scale level encompassing the whole Kehanakorn Village 2 area involving 192 buildings and a surface area of 40,000 sq.meter. All flood adaptation measures are applied at the landscape level, including 12 interventions to be implemented along a pathway with consecutive steps, following a process of top-down adaptation. Fig 4-8(iii) shows the adaptation strategy involving a group of assets following a process of clustered adaptation. For the AR analysis, the EOLC-Gap was set on 10 years and the cluster radius was set on 200 meters. The cluster covers about 42 buildings and encompasses an area of 7,000 sq.meters. The adaptation (OA) commences in the year 2032 with a total of seven options that can be applied to this particular cluster.

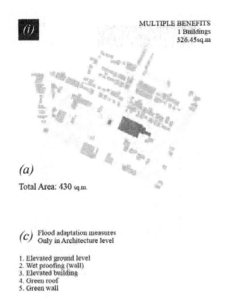

(i)

MULTIPLE BENEFITS
1 Buildings
326.45sq.m

(a)

Total Area: 430 sq.m.

(c) Flood adaptation measures
Only in Architecture level

1. Elevated ground level
2. Wet proofing (wall)
3. Elevated building
4. Green roof
5. Green wall

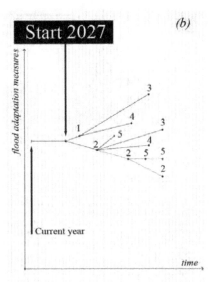

(b)

Start 2027

flood adaptation measures

Current year

time

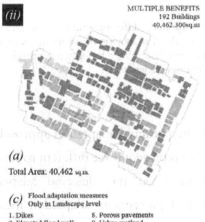

(ii)

MULTIPLE BENEFITS
192 Buildings
40,462.300sq.m

(a)

Total Area: 40,462 sq.m.

(c) Flood adaptation measures
Only in Landscape level

1. Dikes
2. Elevated flood wall
3. Infiltration fields
4. Improve soil
   infiltration capacity
5. Water square
6. Ditches
7. Polder

8. Porous pavements
9. Urban wetland
10. Porous pavements
11. Adding green
    in streetscape
12. Floodway

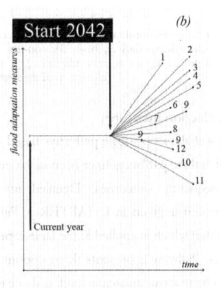

(b)

Start 2042

flood adaptation measures

Current year

time

75

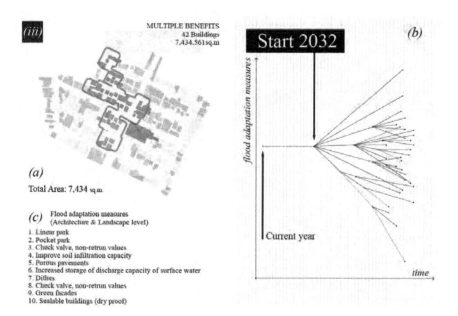

Fig 4-8. The relationship between the decision-tree diagrams at different scale levels: (a) 2-D map (Kahanakorn village 2), (b) the decision-tree, and (c) flood adaptation measures (name) - (i) Individual assets (autonomous adaptation), (ii) whole area (following a top-down approach), and (iii) a group of the assets (clustered adaptation)

4. Selection measures

Fig 4-9 shows possible pathways based on the clustered adaptation approach, of which three pathways have been highlighted to represent three different routes and subsequently outcomes. Detailed information on the clustered adaptation approach is given in CHAPTER 3. Pathway 1 presents an adaptation measure (polder) which is applied at the landscape level and has an estimated lifespan of 27 years. Pathway 2 presents three options with different timing sequences, two of which are at the landscape level, and one is at the architecture level. Finally, Pathway 3 presents three options with different timing sequences, of which all of them are applied at the architecture level.

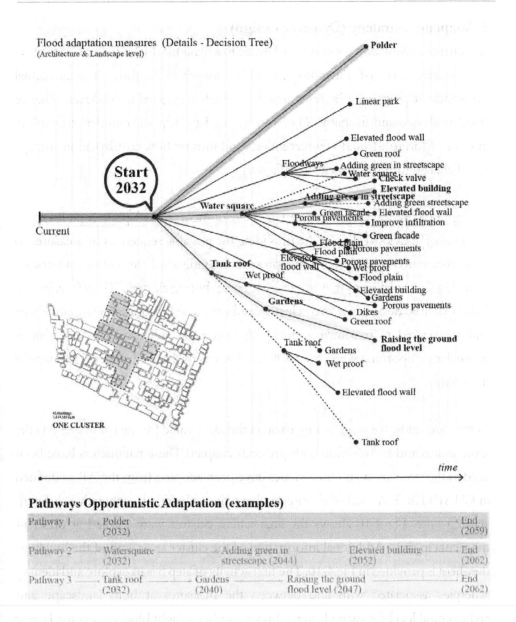

Fig 4-9. The relationship between space and different pathways of the decision-tree diagram

5. Adaptation strategy (Dynamic design)

A sequential adaptation process which leads to a dynamic design for an area is more than merely a set of illustrations, a series snapshots in time. It is an actual development proposal of a transformation which is targeted to gradually enhance flood resilience and livability. They can be used to help stakeholders to explore options of future adaptation interventions and thus on how an adaptation strategy could change the area of a cluster (Fig 4-9).

In step 5 of the methodology (strategy development) the adaptation pathways at the level of individual clusters are providing the possible sequences of measures to be implemented over time. In this process the individual clusters should not be considered in isolation as they interact with neighboring clusters. Therefore, in step 5 the dynamic designs for each cluster are being developed in an integrative way and are steered to maximize synergies between the individual clusters. This is particularly opportune when the EOCLs of assets of these clusters fall in a similar time frame.

In the case study, for neighboring clusters the same value for the parameter has the same radius and EOLC-Gap with previous chapter. These parameters have been used in the simulation and their values have been adopted from the AR as defined in CHAPTER 3. A total of six clusters have been identified, each with a different starting time. Fig 4-10 shows the area where clusters were formed at different momenta in time. Additional information of the cluster areas, and of the buildings they host is provided in Fig 4-10. The first activity of step 5 is to explore and identify synergies associated with and between the measures at both landscape and architectural level for each cluster. Cluster number 4 (light blue) covers the largest area comprising of 58 buildings, 9,000 sq.meter, and starts adaptation in 2031. Cluster number 2 (orange) covers the smallest area (1,700 sq.meter.), including 12 buildings, and starts the adaptation process in the year 2024. These two clusters have the same AR value.

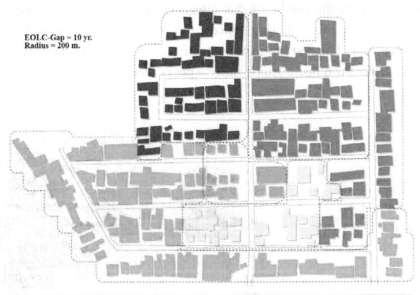

EOLC-Gap = 10 yr.
Radius = 200 m.

Cluster No.	Year	Area (sq.m.)	Unit
Cluster 1	2020	3,009.5	20
Cluster 2	2024	1,706.1	12
Cluster 3	2027	3,102.5	20
Cluster 4	2031	9,072.6	58
Cluster 5	2034	7,434.1	42
Cluster 6	2039	7,364.5	40
Total		32,689.3	192

Fig 4-10. Clustered adaptation strategies

The resulting pathway for Cluster 1 is described in detail in Appendix C (as an example). Fig 4-10 depicts the selection of the neighboring clusters (six clusters) including the EOLC which marks the start of the adaptation pathways of each individual cluster.

Fig 4-11 presents the results of this integrated, dynamic design process. The subfigures depict the following aspects: (a) spatial occupation and (b) the perspective view. The first image presents the current situation, the other images (designs) depict the transformation process per cluster.

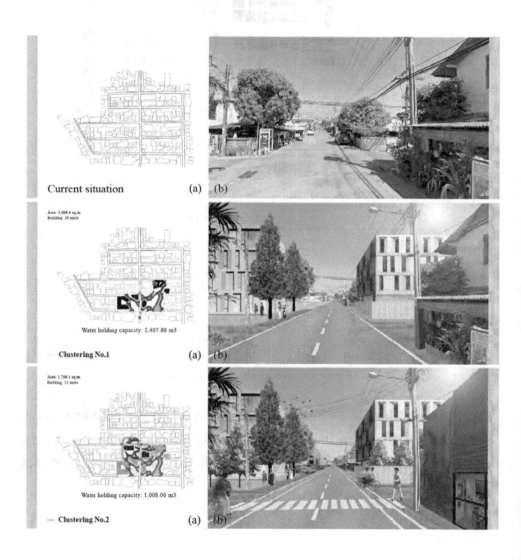

Current situation   (a)   (b)

Area: 3,008.4 sq.m.
Building: 20 units

Water holding capacity: 2,407.60 m3

Clustering No.1   (a)   (b)

Area: 1,708.1 sq.m.
Building: 12 units

Water holding capacity: 1,008.00 m3

— **Clustering No.2**   (a)   (b)

Fig 4-11. Example perspectives (all clusters)

It follows from the simulation and the dynamic design process that as of 2031 (cluster no. 4), the highest number of buildings that share common benefits is 58 units covering about 9,000 sq.meter and providing a capacity for water storage of 7,000 sq.meter. The average size of the area used for storing water is 31,200 sq.meter, and the required time to install this capacity is fourteen years.

4.4 Discussion and conclusions

The integration of water management in the urban design requires a design-based approach in which solutions are sought to go beyond the primary objective. Water is not purely utilitarian, but actually adds to the urban quality which is reflected in the actual building program at block or neighborhood level. This case study has demonstrated multiple adaptation pathways to exist and to which co-benefits have been added to and amplified in the urban landscape. Hence, adapting to climate change provides a window of opportunity to simultaneously improve the quality and livability of urban areas. Unlocking these opportunities will require a carefully designed planning process in which the potential pathways need to be identified and compared.

An extensive catalog of adaptation measures for different scale-levels ensures flexibility in selecting the type of measures that can be integrated. The case study revealed that sequencing adaptation measures over longer periods of time (adaptation pathways) provided insights into the direction of the adaptation strategy to the area which in turn can be achieved and managed. Nature-based solutions are playing a pivotal role in this process and aspiration.

The number of available adaptation options usually decreases with time when the landscape transforms and, as a consequence, as time passes by, reduces the number of options to create additional value (co-benefits). In other words, options to enhance the livability and quality of life of an urban area at a future point in time will be smaller than those potentially available on the short-term. This observation supports the importance of an integrative approach and of seizing the opportunity to adapt when chances to intervene are becoming available. Integrating adaptation measures at the architectural level into the landscape level increases the number of options and thus the potential for integrating more options to unlock multiple benefits. In other words, seen from the perspective of the decision tree, choosing a measure which is applied at the architectural level limits further choices to be implemented at the same level in a later stage. A measure applied at the landscape level generates more freedom to choose, as both landscape and architectural measures are still available in the future. The degree of flexibility in the future can be managed by following a certain pathway in the decision tree taking into account the options potentially available on the longer term.

This case study following a dynamic design approach has revealed some specific new insights and lessons with regards to the developed methodology. These are:

- *Combining clusters.* In the real world, a clustered, integrative and dynamic design approach based on EOCL does not fit into the current building programs, which lack flexibility from a view point of temporal as well as spatial perspectives. This notion stresses the importance of long-term strategic planning to maximize the exploitation of synergies between adaptation interventions. Only then, when time horizons of at least a few decades are being adopted, and the dynamics of autonomous renewal cycles are understood, opportunities will emerge which allow to lift the focus of a standalone intervention at the scale of individual assets, to the scale of a cluster or even beyond, to the scale of a combination of clusters. This particularly has relevance for green infrastructure projects which are often of too little scale and too fragmented to yield significant impact. The proposed OA approach has potentials to reverse this trend.

- *Design-synergy with building programs.* The design of flood resilient neighborhoods in which flood prevention is integrated in all aspects of buildings, infrastructure, public space and urban vegetation & surface water systems provides extensive design opportunities for fundamentally changing the character of the area. When only applied at building level, these opportunities are far more limited and might even result in maladaptation due to increased fragmentation.

- *Co-benefits*. Adapting larger areas including infrastructure, public space and existing urban vegetation/surface water bodies provide ample opportunities to create and add co-benefits. Livability, health, property value, ecology and many other aspects are enhanced from an integrated adaptation approach encompassing areas in which typically nature-based solutions are applied at landscape level.

CHAPTER 5
Decision tree method for evaluating the flexibility of flood risk adaptation options

Operational method	
	CHAPTER 2

Assessment and analysis		
CHAPTER 3		CHAPTER 4
Assess adaptation potential	Identify adaptation options	Assess current and future flood risk
Lifecycle assessment methodology	Catalogue and classification	Flood modelling/empirical data (beyond scope of this study)

Developing adaptation pathways		
CHAPTER 3 + CHAPTER 4	CHAPTER 3 + CHAPTER 4	CHAPTER 4
Define clusters	Map opportunities	Identify pathways
Urban density, costs/service-benefits	Selection of options	Decision Tree method

Towards a strategy	
CHAPTER 5	CHAPTER 6
Evaluating flexibility	Governance
FIA method (measuring flexibility)	Interviews and survey, Ex ante evaluation

This chapter is based on the journal article "Nilubon, P., Veerbeek, W. and Zevenbergen, C., 2019. Decision Tree Method (Flexibility in Adaptation–FIA) for Evaluating the Flexibility of Flood Risk Adaptation Options in Lad Krabang, Thailand. *International Journal of Water Resources Engineering*, 5(1), pp.32-48."

5.1 Introduction

To develop, analyze and evaluate climate adaptation policies, strategies and measures, there are multiple methods available. A key feature of most of these methods is that they allow to evaluate their flexibility. Examples of such approaches are Adaptation Pathways (AP) and Dynamic Adaptive Policy Pathways (DAPP) (Haasnoot, 2013, Gersonius et al., 2013; Woodward et al., 2014; Meyer, 2018; Gersonius, 2012; Wise et al., 2014; Byg and Herslund, 2016). In this chapter these methods are captured by the umbrella term "Adaptive Pathways"[2]. Flexibility is defined here as the ability (i) to change a policy, strategy or measure and/or (ii) to increase or slow down the implementation or execution (rate) of a policy, strategy or measure. It follows from the above that these methods aim to identify potential future options, but avoid becoming "locked-in" or creating a singular intervention or approach (policy, strategy or measure). Adaptive pathways have also in common that decisions to change a strategy (or replace an adaptation measure), referred to as decision points or tipping points, are typically triggered by a predefined and single driver related to environmental or social change which can be monitored and ideally show a trend. The method developed in this study, referred to as OA using Adaptation Pathways, defines tipping points as momenta in time to intervene or change the strategy triggered by 'an opportunity' emerging from autonomous urban renewal processes instead of by a signal which refers to a gradual environmental or social change. In other words, the method developed in this study is about

---

[2] The use of the terms 'adaptation' and 'adaptive' is as follows.
Adaptation is used in two different ways. In the context of climate change, it refers to measures that aim to prepare for the consequences of a changing climate. The alternative use of the term refers to an adjustment of a plan or strategy following new insights. The adjective 'adaptive' is associated with the second meaning of adaptation and often refers to a plan which is not based on static conditions and takes uncertainty into account.

providing guidance on 'Adapting when we *can*' instead of 'Adapting when we *have to*'.

Adaptive Pathways are widely used in FRM and water supply management (Haasnoot et al., 2012; Haasnoot et al., 2013; Eichler et al., 2015; Ren et al., 2011; Buurman and Babovic, 2016; Zandvoort et al., 2017; Ontl et al., 2017). However, it still remains a challenge to use these methods in practice and studies which evaluate adaptation pathways methods are scarce in these domains. Studies in The Netherlands have shown that Adaptive Pathways have been instrumental to bring stakeholders together in the initial phase of the development of the national FRM strategy referred to as the Dutch Delta Plan, to create awareness about uncertainties in future projections and to arrive at a shared, common notion among the stakeholders of the importance of long-term planning to manage flood risks. Difficulties in the timely detection of tipping points due to the large natural variability of the signal values of key drivers are, amongst others, hampering its application in the implementation phase of the strategy (Bloemen at al., 2016).

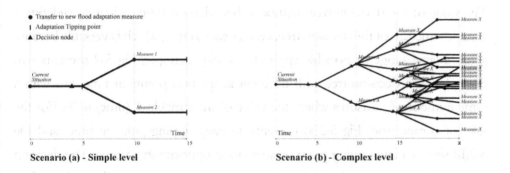

Fig 5-1. Decision-tree diagram presents a simple and complex level

In this study, a Decision Tree Method has been developed to provide insight into the sequence of measures to be considered for implementation based upon adaptation opportunities (note: sequence of measures refers in this context to an adaptation pathway). It uses the EOLC as the guiding principle to replace or adapt the measure or strategy. By contrast, adaptive pathways methods are generally based upon climate change scenarios using (amongst others) sea level rise and river discharge levels to indicate when a measure or strategy need to be changed.

By its nature, a Decision-Tree method enhances the FRM approach, as it considers every chance to adapt as an opportunity to strengthen the system. A decision tree combines both an external driver, such as flood conditions and an internal driver, the lifespan of flood adaptation measures, to generate pathways. A decision tree algorithm potentially generates more than one pathway, with the possibility to model this in software - in order to automate the process of evaluation (see also CHAPTER 3). These calculations can produce results with widely varying levels of complexity, from very simple (See Fig 5-1(a)) to very complex (See Fig 5-1(b)) outcomes. The advantage of computer calculations is that they create and analyze a broad spectrum of possible outcomes in a relatively short period of time. The more time is allocated for the calculation, the more measures can be processed and thus the more complex the outcome will be.

The value of the decision-tree method is best demonstrated when considering a generic structure of the decision tree, i.e. an ordered tree, which is versatile because it has a large range of possible applications. For example, Fig 5-2 presents two examples of a decision-tree with different adaptation points in time, but with a same number of measures where the value of the variable T (time) is 20. For the sake of comparison, Fig 5-2(a) presents an early tipping point in time, and Fig 5-2(b) shows a late tipping point or adaptation opportunity in time. The decision to choose between an early or a late adaptation intervention is one that is dependent upon (amongst others) the urgency to take action, the available resources and the level of uncertainty in the prediction of future flood risks of the affected area. The higher the uncertainty, the higher the value added of postponing an intervention and having adaptation options available at a later point in time. Hence, uncertainty calls for flexibility. The question arises of how to value flexibility and how to take this value into account when developing a decision tree? The Decision-Tree

Method in its current version lacks the possibility to weight flexibility and to assign an absolute value to the degree of flexibility with respect to the adaptation measures.

In this chapter a Decision Tree Method (as described in CHAPTER 4) will be extended to include flexibility as a criterion to identify and select adaptation options. The objective of the proposed modification of the method is to develop a sequence (or sequences) of adaptation measures in time (e.g., adaptation pathways) in which flexibility has been factored in into the weighting of the adaptation measures. Hence, it enables to include flexibility as a criterion in the ranking and sequencing of measures in time.

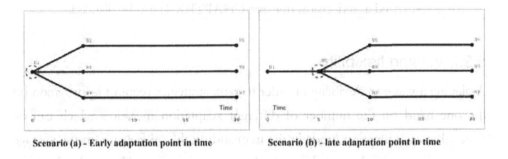

Scenario (a) - Early adaptation point in time   Scenario (b) - late adaptation point in time

Fig 5-2. Decision-tree shows the action point in time coined as early and late adaptation

The approach which allows to introduce flexibility into the development of adaptation strategies, is by comparing the characteristics of different pathways of a decision tree and weight their relative flexibility. The steps proposed to identify, rank and select pathways combine the decision tree diagram with the LCA method (see for instance Rossi et al., 2012; Zheng et al., 2014; Nilubon et al., 2016) and compare the different pathways on the basis of their flexibility.

5.2 Proposed modification of the methodology (to include flexibility in the Decision-Tree Method)

5.2.1 Selecting indicators of flexibility

The decision tree characteristics encompass two features which are equally important for valuing flexibility. These are (i) the node distance and (ii) the node branching factor (Patil et al., 2010). The node distance represents the period of time where no measures can be applied, hence from a flexibility perspective, this value should be as low as possible. The other feature, the node branching factor, defines how many adaptation options are available at a particular moment in time. This value should be as high as possible as to maximize flexibility of the adaptation strategy. Based on those two features, indicators to value the flexibility of a strategy (which is visualized as a decision tree, see CHAPTER 4) can be distilled.

5.2.2 Valuing flexibility

Assigning a value to flexibility in order to rank strategies requires information on the one hand on the number of different adaptation measures which can be selected at each junction (node in the tree) and on the other hand on the average time required to implement the next adaptation measure(s). Hence, the flexibility of an adaptation strategy is positively correlated with the number of adaptation pathways. The following mathematical formula allows to combine these indicators and to arrive at one (normalized) value of flexibility for an adaptation pathway (N_n):

$$N_n = \left(\frac{B}{B_{max}} \right) + \left(\frac{De_{max}}{(De + De_{max})} \right) + \left(\frac{Di_{max}}{(Di_{avg} + Di_{max})} \right) + \left(\frac{Di_{avg}}{Di_{max}} \right)$$

--- (Eq 1)

To normalize the outcomes the indicators are expressed as fractions of the normal values (see Section 4.2.1) and the maximum values: B_{max}, De_{max} and Di_{max} respectively (Eq 1) in order to be able to compare one factor to another. These values are precomputed and allows to compare different decision trees:

1. The first term of the formula expresses the node branching factor divided by a maximum node branching factor. Obviously, this term gives a high score when the branching factor is high.

2. The second term of the formula relates to the lifespan of measures. As explained previously we want to minimize tree depth at the current node, in order to obtain a higher potential for adaptation. The value of this term starts at 1, for a tree with depth, and decreases towards 0, as the depth of the tree becomes larger.

3. The third term relates to the requirement of staying agile and flexible, giving a higher score to measures that do not take long to be implemented. While the previous term was focused on tree depth, this one is trying to minimize the largest distance between any two nodes (i.e. measure implementation time).

4. The last term has a similar purpose as the previous one, but instead of focusing on the largest distance between two nodes, it minimizes the average distance between all connected nodes in the tree. This term is designed to differentiate between similar but not identical trees, where the previous factor is sufficient to produce a better or worse flexibility value

The four factors together are expressed in Eq 1 which computes a score N_n that reflects the value of flexibility of an adaptation pathway of a strategy which is presented as a decision-tree. The flexibility score of the strategy is calculated by taking the sum of the flexibility scores (N_1 to N_n) of all the adaptation pathways of the decision tree. This is referred to as "the average flexibility of an area (FIA)":

$$\text{FIA} \ = \ \frac{N_1 + N_2 + \ldots N_n}{n} \quad \text{--- (Eq 2)}$$

As mentioned, depending on the local institutional and socio-economic context (such as stakeholders urgency to intervene or wish to postpone capital intensive investments, the degree on uncertainties in future projections), it could be desirable to favour early adaptation or instead the opposite, to favour late adaptation. Early adaptation implies a strategy with a disproportionate number of nodes with high branching factors relatively early in the tree. To have the option (flexibility) of early adaptation may lead to a preference for a particular strategy or even a selection of particular pathways of a strategy. Hence, this type of flexibility needs to be taken into account when developing an adaptation strategy.

In order to consider and include early adaptation as a criterion of flexibility for the development of an adaptation strategy, two factors have to be taken into account. These are (i) the time it will take to implement the first measure (N_1) and (ii) the number of measures to be implemented on the short term or the node branching factor (B_t) relative to the total number of measures or the maximum node branching factor (B_{max}) of the strategy. As it is common practice to consider 10 years as a short-term time horizon in urban planning (see for example Bloemen et al., 2018), in this study we adopted this short-term horizon to calculate the values for this factor (B_{10yrs}/B_{max}).

It follows from the above that in addition to FIA, indicators related to the timing of interventions and the number of interventions have to be included to arrive at a meaningful, encompassing assessment of the flexibility of an adaptation strategy. In Table 5-1 an overview of these indicators is presented as well as the equations of how to quantify each of these indicators. To allow easy intercomparison, the values are normalized to a range between 0 and 1. These indicators comprise:

1. The potential measures available. The first equation computes the total number of potential measures.
2. The total number of adaptation pathways. The second equation computes the total number of adaptation pathways.
3. FIA. The third equation computes the value of FIA without concern of the

intervention timing (early and late adaptation).

4. The lead time (of the first measures). The fourth section equation concerns the lead time of the first measures implemented.

5. The number of measures on the short term. The fifth equation computes the number of measures in the short term (t) implementation.

6. FIA(t). The sixth equation computes the value of the FIA concerned with the timing of the intervention (early and late adaptation).

Table 5-1. Indicators and equation to assess the flexibility of an adaptation strategy.

Indicators	Equation	
1. The potential measures	$$\left(\frac{B}{B_{max}} \right)$$	B (the number of measures in time) B_{max} (the total number of measures)
2. The adaptation pathways	$$\left(\frac{De_{max}}{(De + De_{max})} \right) + \left(\frac{Di_{max}}{(Di_{avg} + Di_{max})} \right) + \left(\frac{Di_{avg}}{Di_{max}} \right)$$	De (time for implementation), and Di (lifespan of measures), De_{max} (the total years of the decision tree), and Di_{max} (longest lifespan)
3. FIA (flexibility in adaptation)	$$\frac{N_1 + N_2 + ... N_n}{n}$$	N (the flexibility for each strategy), and n (the total amount of strategy)
4. Lead time for the first implementation of the measure	$$\left(\frac{t}{t_{(N_1)}} \right)$$	t (time for applying the short-term strategy), and $t_{(N1)}$ (first implementation period)
5. The number of measures in the short term (t) implementation	$$\left(\frac{B_t}{B_{max}} \right)$$	B_t (the number of measures to be implemented in the short term (t) or the node branching factor)
6. FIA(t) intervention of timing	$$\left(\frac{B_t}{B_{max}} \right) + \left(\frac{t}{t_{(N_1)}} \right)$$	

5.3 Applying the methodology in Kehanakorn village 2, Lad Krabang

The methodology described in 5.2 has been applied in a case study area (see also CHAPTER 4). The case study area comprises a section (neighborhood) of Lad Krabang, Kehanakorn, servicing 192 buildings distributed over a total area of 40,460 sq. meters. At the heart of this methodology is the decision tree, which presents all possible measures applied at different points in time and thus encompasses strategies of which each of them follows a certain pathway. The design of the decision tree is based upon computer calculations (using a model (see Appendix D) with the parameters: the potential measures (from the catalog), lifespan of measures, area size, flood inundation depth, etc). In CHAPTER 4 a detailed description of the model is given as well as the input data and assumptions (4 strategies) used in this case study. The indicators and corresponding equations of Table 5-1 have been used to quantify the flexibility of the resulting adaptation strategies. These values of flexibility are also used to compare the different decision trees representing a collection of multiple strategies (options).

The weight values which were used to assess the flexibility of the decision trees, were calculated using the results of the case study area as presented in CHAPTER 3 and CHAPTER 4. Simulations with the model have been carried out for four adaptation strategies comprising:

1. Autonomous adaptation
2. Top-down adaptation
3. Clustered adaptation, including *2 variants*: (1) r 200 and EOLC-Gap 5 and (2) r 200 and EOLC-Gap 10

Fig 5-3 presents the results of the model calculations. They show the relationship between flood adaptation measures (decision-tree) and the size of the areas for a selected adaptation strategy. Fig 5-3(a) presents the results of the "autonomous adaptation" strategy for an area of 450 sq. meter following from the value of the end of the life cycle of the buildings which has been used as input for the model calculations. The resulting decision-tree includes three flood adaptation measures at the architecture level. These are: Tank Roof, Green Roof, and Gardens. The first adaptation intervention starts in the year 2027.

Fig 5-3(b) presents the results of "Top-Down Adaptation strategy". The area's size is 40,460 sq. meters, the number of buildings is 108, and the size of the (public) area where multiple benefits are shared is 15,200 sq. meters. The first adaptation intervention starts in 2042.

Fig 5-3(c) and 5-3(d) show the results of the "Clustered Adaptation" strategy for two combinations of values representing two clusters. The model outcomes provide two groups of assets over a radius (r) of 200 meters, with a lifecycle gap (g) of 5 years and 10 years, respectively. The first group (Fig 5-3(c)) with a total area of 4,526 sq. meters, includes 20 buildings which occupy 3,009 sq. meters hosting citizens and other local stakeholders who benefit jointly from the local services provided by the public spaces in that area. The first adaptation intervention starts in the year 2020.

The second group (Fig 5-3(d)) has a radius r for grouping assets of 200 meters and a total area of 9,485 sq. meters. A total of 42 buildings have been identified situated in a combined area of 6,254 sq. meters. The citizens and other local stakeholders share some benefits. The first adaptation intervention starts in the year 2032.

The results of the flexibility assessment are presented in Fig 5-4 as a radar chart capturing the six flexibility indicators of the four adaptation strategies. From this chart it can be observed that both the top-down as well as the bottom-up (autonomous) adaptation strategy have relatively low flexibility values when compated to the clustered adaptation. For the bottom-up strategy this indicator is the number of potential measures and for the top-down strategy this indicator is the lead time. The latter clearly indicates that the autonomous adaptation strategy comprises small scale measures which relatively short lead times. This substantiates the general feature of autonomous adaptation strategies to have the ability to intervene and adapt relatively fast as opposed to top-down adaptation. In addition, the chart also reveals that the two clustered adaptation strategies have relatively high values for each of the six indicators, which support the assumption that clustered adaptation enhances the flexibility of the strategy.

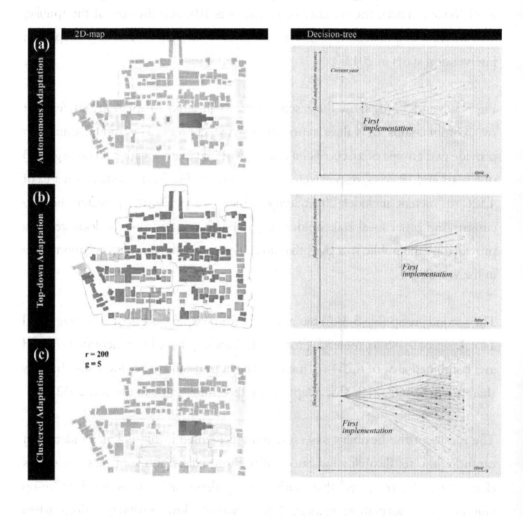

Fig 5-3. The relationship between flood adaptation measures (decision-tree) and size of the area
which represents the value of flexibility of the adaptation strategy (autonomous, top-down, and
clustered)

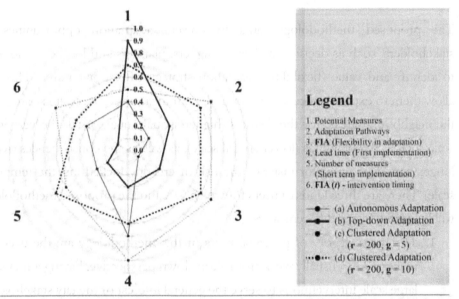

Strategy	Year of First implementation	Area (sq.m.)	The number of buildings
(a) Autonomous Adaptation	2027	430.5	1
(b) Top-down Adaptation	2042	40,462.3	108
(c) Clustered Adaptation (r = 200, g = 5)	2020	4,526.9	20
(d) Clustered Adaptation (r = 200, g = 10)	2032	9,485.1	42

Fig 5-4. Comparison of the performance of key features (represents the value of flexibility in the
decision tree) of the four adaptation strategies (autonomous, top-down, and clustered). The values
are between 0 and 1, expressed as an index. Dotted and dashed lines represent the results of the
clustered adaptation strategy (c) and (d), respectively, while the thick and light line represents the
results of the top-down and autonomous adaptation strategy, respectively

5.4 Reflection on the methodology

This study is one of the first to develop and apply a quantitative methodology to assess flexibility as a multi-component parameter in designing an adaptation strategy based on the principle of OA. The selection of parameters is subjective and has been made based on common sense and discussions with stakeholders (see CHAPTER 6). In this study, a pragmatic approach has been followed. Flexibility has been framed as the capability of a strategy to change course and steep up or slow down the rate of implementation depending on the needs, changing local conditions, and the arrival of new information about future projections.

The proposed methodology aims to present adaptation opportunities to stakeholders, such as decision-makers, designers, planners and local communities to identify and value (flexibility) adaptation strategies for a given area. This will allow them to explicitly take into account the ramifications of a certain strategy for the neighboring areas, for the districts (higher spatial scale), or for the very local scale such as the buildings (lower spatial scale), as well as the consequences over a longer period of time (for instance the risk of creating locked-ins, the temporal scale). There are three kinds of users for which OA and the proposed methodology will be useful, but in different ways:

1. The first category of potential users of this methodology are the decision makers. They usually work from a top-down perspective, having a focus on large scale interventions to serve the general interest of the city stakeholders and to comply with national flood standards. For some decision makers, the time horizon is relatively short (next 5 years), as they have to show impact, but others may have a longer time horizon (even more than 50 years) depending on their political stance and commitment and where they sit in the city government. This methodology will offer the decision makers the information required: what are the options on the short term? What are the consequences of selection a strategy and how flexible is my strategy (does it allow any adjustments on the longer term, when do we need to make the

decisions on large investments?)

2. The second category of users is the local community. The local community needs to be timely informed about the consequences they face due to adaptation actions taken by the city government and their underlying assumptions. It will be in the interest of all stakeholders to engage the local communities in the process of decision making as their ability to contribute and mitigate urban flood risk is more and more recognized. Till date, however, their local interventions often lack coordination and application in an effective way. The result of this methodology aims to support effective engagement of the local communities in this process of decision making.

3. The last category of users (i.e. designers) are architects and engineers. They will become more and more facilitators in the process of decision making. They are becoming increasingly aware that there is no single and best solution for addressing the flood risks cities are facing now and in the future. Providing a broad range of options and consequences to the stakeholders, showing where there can act and what the consequences of their action are, will become crucial in the near future. This methodology aims to deliver this information and therefor it is assumed it will be a useful tool particularly for designers and architects in their new role.

5.5 Conclusions

This chapter presents a new methodology to identify and value flexibility of an adaptation strategy based on decision trees and adaptation pathways. This methodology, FIA, has been developed to evaluate and, most importantly, to offer a comparison between adaptation pathways, which are generated by the decision-tree method. FIA also considers a raft of indicators, such as the timing of interventions, the number of potential measures, and the lead time of implementation, to value flexibility for developing an adaptation strategy. The application of the methodology in the Kehanakorn village 2, Lad Krabang district, showed that a clustered adaptation approach, where interventions at the lowest spatial scale are integrated and being considered at a higher spatial scale in a coordinated way, resulted in the highest values for flexibility, as opposed to the two alternative approaches (e.g., top-down and a bottom-up). It is expected that this methodology to identify the various adaptation options (pathways) potentially available together with information about the flexibility (and thus partly on the strategic consequences of options), will support stakeholders (including designers and architects) to actively engage in a joint process of designing an adaptation strategy.

CHAPTER 6
Challenges to implementation of an Opportunistic Adaptation approach in Bangkok

This chapter is based on the journal article "Nilubon, P., Veerbeek, W., and Zevenbergen, C. (2019) Challenges to implementation of an Opportunistic Adaptation approach in Bangkok, *International Journal of Water Resources Engineering*, 5(2), 28-45."

6.1 Introduction

Bangkok, the capital of Thailand, is the seat of the national government which is in charge of making planning policies for large infrastructure projects such as flood protection infrastructure (Thanvisitthpon and Shrestha 2018). While the national government is the decision-maker at the central level, the Bangkok Metropolitan Administration (BMA) is responsible for the formulation and implementation of (local) planning policies, including managing the flood risk in the Bangkok Metropolitan Area. It has a special mandate and responsibility for the public safety and security within the jurisdiction of Bangkok.

The BMA has responded to the catastrophic flood event of 2011 with a 10-year 'Master Plan on Climate Change' 2013 -2023 (Fujimoto et al. 2016). It covers both adaptation and mitigation, but its implementation has been delayed due to political reasons (Marks 2015; Marks and Lebel 2016). Aside from this plan, there are other national initiatives as a response to the 2011 flooding, like for instance the construction of a dam in the Nakhon Sawan province, located upstream in the northern part of the country. According to the national government, this mega-investment would significantly reduce flood levels in downstream areas. Yet, experts argue that this would only reduce the water level in Bangkok by about 10 cm. This is based on flood data gathered in 2011 (Ogle 2013). The lack of agreement among experts has been the cause that the project has been eventually abandoned (Marks 2015). By and large, the BMA has primarily focused on fluvial flood mitigation measures, such as dikes, along the river and canals (i.e. large and medium scale flood protection measures). While these measures can prevent much of the annual flooding from rivers and canals, they do not effectively mitigate pluvial flooding in urban areas. Indeed, recent experiences demonstrate that small scale pluvial flooding continues to occur. Moreover, according to Meehan (2012) excessive rainfall was the primary cause of the 2011 flooding and not river overflow, as the first comprised 75% of the total flood water in November 2011.

Frequent and small-scale pluvial flooding forced many homeowners to adapt their assets by themselves. The sparse flood prevention measures that were implemented by the local government agencies were planned without much consultation and involvement of the residents. These interventions included small water pumps, which were placed at various points across their neighborhoods to cope with heavy monsoon rain. Other measures implemented by the local government (the Lad Krabang District Office) included cleaning the drainage system and retention ponds (McGrath and Tachakitkachorn 2013; Saito 2014), maintenance of canals, and enforcements of dike systems. However, these measures proved neither to be practical, nor provided a sustainable long-term solution. Additionally, there are communication and management problems related to the internal governance structure. For example, the Department of City Planning and the Department of Drainage and Sewerage do not exchange information effectively during periods of intense rainfall (Nuchudom and Fongsuwan 2015). In addition, the national government has always resorted to reactive policies as a way of coping with flood hazards. As a consequence, it has focused on short-term planning only. Long-term strategies and plans have hardly received attention.

The need for adequate flood management arrangements is especially urgent in poor urban communities as flooding may lead to chronic diseases and thus may exacerbate poverty-related diseases (Adelekan 2010). These diseases are driven (amongst others) by potable water shortages caused by water pollution and damage to water pipes following flood events. Several studies argue that the best approach to increase the level of resilience in poor urban communities is to scale up adaptation (i.e. from individual to neighborhood level) to effectively reduce flood risk (Muller 2007; McGranahan et al. 2007; Dodman and Satterthwaite 2009; Jabeen et al. 2010). Autonomous adaptation (i.e. unplanned individual adaptation) often involves interventions to mitigate impacts of flooding at the household level. At a higher spatial level, these measures may be inefficient or inadequate. Autonomous adaptation will benefit those who take action. Yet they can transfer the flood risk to others and become maladaptive (Limthongsakul et al. 2017).

Unfair risk distribution caused by autonomous adaptation, often creates tensions and conflicts among stakeholders. Taken all these adverse consequences together, autonomous adaptation may increase local flood risk affecting the whole community (Tanner et al. 2009; Mycoo 2014; Sin-ampol et al. 2019).

In this dissertation it is assumed that OA will ignite and ease the alignment between government-driven top-down flood adaptation and bottom-up driven initiates developed by homeowners and other local stakeholders (see CHAPTER 3). However, this assumption has not yet been appraised (based upon expert judgments) nor verified (applied in a concrete context).

Roundtable discussions with different stakeholder groups and experts in this field have been organized to evaluate this assumption. In this chapter, the outcomes of these discussions will be presented. In addition to these roundtable discussions, a series of interviews, as well as a survey among residents of Lad Krabang, have been conducted (see Section 6.4). A central theme in the roundtables has been the flooding of 2011 and its aftermath. The roundtables focused on identifying the lessons learned, particularly related to governance processes underlying the selection and implementation of responses (autonomous and planned) as well as to the proposed longer-term strategy.

The roundtables were organized in Bangkok in the period from July to August 2018. The three main goals were as follows:

1. To reflect on the current FRM strategy and in particular on its governance aspects;
2. To assess stakeholder's perception of the potentials of the OA approach;
3. To distill (based on outcomes goals 1 and 2) conditions required to implement the OA approach.

6.2 The governance of flood risk management in Bangkok

The Thai Disaster Prevention and Mitigation Act (2015) encompasses a description and analysis of a disaster management arrangement that covers all types of disasters, policy guidelines, operating procedures and coordinating procedures. This act addresses both the identification and maintenance of flood protection through design standards and post-disaster risk response (DRR) in the short-term as well as in the long-term. For example, it addresses regulation criteria for flood mitigation including financial subsidy arrangements.

The organizational structure of the national government department (see Fig 6-1) comprises the following three levels:

1. National level. At this level, it is the responsibility of the National Civil Defence Committee (NCDC) to coordinate all activities relevant to civil defence and disaster management, as well as post-disaster response (DRR). This includes the activities performed by the Office of National Water Resources (ONWR), the National Safety Council of Thailand (NSCT) and the Department of Disaster Prevention and Mitigation (DDPM). This committee, acting as the head of the other departments, is chaired by the Minister of Interior, and the membership comprises representatives from 34 relevant national government agencies.

   A misconception, which is often observed in practice, is that the NCDC is solely responsible for the management of disasters and for facilitating disaster recovery. However, an essential mandate of the committee is also to organize disaster prevention. This includes maintenance and upgrading of all flood defense works in the country. It should be noted that urban drainage-related issues are typically managed by the municipality.

2. Regional level. At the regional level, 12 Regional Disaster Prevention and Mitigation Centers have been established to render technical assistance and

auxiliary services to local Civil Defense Committees. At the level of the central government coordination between the regional centers will take place: the regional governors receive information from the central government only.

3. Local level. The main actor who is in charge of the execution of actions in the event of disasters is the local Civil Defense Committee in both rural and urban areas. The cities of Bangkok (Bangkok Metropolitan Area, and Pattaya City) have dedicated agencies responsible for planning, execution, and maintenance of flood mitigation structures. For the Bangkok Metropolitan Area it is the previously mentioned BMA respectively.

As far as the BMA is concerned, the Department of Drainage and Sewerage (DDS), which sits in BMA, is responsible for the management of the flood mitigation systems. The management is driven by design standards which in turn guide the planning and design of these systems. Hence, BMA is providing technical support. Although the jurisdiction of BMA is the Bangkok Metropolitan Area, it also offers technical assistance outside the Bangkok Metropolitan Area.

From a global perspective, it is increasingly accepted that flood protection and management require all stakeholders to play their given roles. In other words, flood management is more and more conceived as a joint responsibility and not the responsibility of a single governmental organization anymore (van der Werff 2004; Edelenbos et al. 2017; Lamond et al. 2019; Van Der Plank et al. 2019). Hence, cooperation and collaboration among all concerned governmental and non-governmental agencies are indispensable (Palttala et al. 2012).

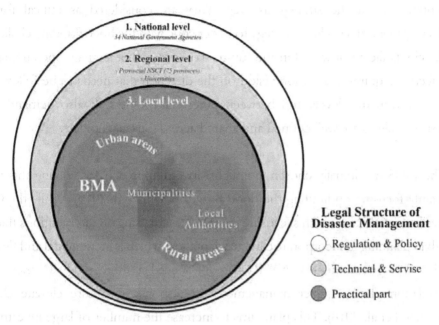

Fig 6-1. Organization structure diagram of disaster management of Thailand (2015)[3]

Bangkok is a member of the Rockefeller Foundation's 100 Resilient Cities[4] project (Laeni et al. 2019), which aims to better equip member cities to mainstream urban climate resilience in their daily practice. Along these lines, the Thai government has embraced 'resilience' as the main goal for Bangkok. According to Harguchi and Lall, 2015, the physical interventions identified to accomplish this goal have been selected and prioritized by various experts and BMA officials. The interventions proposed included amongst others: redistribution of water flow, installation of more water storage areas (such as Monkey Cheeks), and construction of infrastructure that prevent retained water from flowing into neighboring areas. In addition, creating awareness and knowledge sharing between both the governmental organizations and other stakeholders have been prioritized as a

---

[3] http://www.disaster.go.th/upload/download/file_attach/584115d64fcee.pdf

[4] Currently referred to as the Global Resilient Cities Network (GRCN)

central part of the strategy as well. They are considered as critical for the development of a resilient strategy to better cope with the huge flooding challenges the city is facing now and in the future. To have such a strategy operational and effective requires a common vision on the direction that needs to be followed at all governmental levels and between other stakeholders. It also demands that responsibilities are well defined and shared across the stakeholders.

The DDS has initially chosen a straightforward approach for solving this issue, mainly focusing on limiting the flood hazard (Laeni et al. 2019). While the Chief Resilience Officer (which sits inside the BMA) is focusing on interventions that use urban planning to incorporate climate adaptation, which is incremental and flexible. The Thai government (ONWR) has announced a more rigid 20 year 'water regulation plan' (disaster management) in response to future climate change (Richaud et al. 2018). This plan aims to increase the number of large investments of flood protection measures such as giant tunnels and flood walls along the rivers and canals to reduce the risks of future extreme weather events (Wongsa et al. 2019). These two approaches of flood mitigation, namely (i)'climate adaptation' with a focus on long-term, incremental change and flexibility and (ii) 'disaster management' with a focus on preserving the status quo and robustness, seem hard to reconcile in terms of governance as the responsible governmental institutions work in silos, as separated organizations on separate thematic priorities. For example, the focus of BMA's activities is on management and planning in accordance with the 20-year water regulation plan set by the Thai government. This water regulation plan is rigid and allows engagement of the public (and other stakeholders) in the plan making phase only to a limited extend. Finally, the disaster action plans and policies, which are determined by ONWR, are based on a single scenario, and are not designed to allow adjustments if future conditions change.

6.3 Interviews and survey

To assess the perception of stakeholders on the current flood management practices and the potential benefits of OA, unstructured interviews and a questionnaire supported by the workshop were held in July 2018. The unstructured interviews and the questionnaire method aim to evaluate the feasibility of the OA approach from a governance perspective. 'Unstructured interviews' which were carried out, comprised of questions of informal nature and open ended. Generally, this resulted in harvesting more information than expected. The target audience was mainly composed of staff from the institute of Metropolitan Development as well as from the local authorities and representatives from the local community. The staff of the Institute of Metropolitan Development included national government officials from the ONWR, NSCT, DDPM, as well as from the BMA. Representing the local authorities, involved the local government such as the Lad Krabang District Office, as well as the local community. The main objective of the interviews was to assess the enabling conditions of the proposed OA approach and to identify potential constraining factors.

The assessment methodology has been divided into two parts based on the questions that were used. The first part covers the unstructured interviews and the second part the questionnaire. Both focused on two main subjects: 1) the current flood management practice, and 2) the OA approach. The questions were targeted at two main groups: the experts (e.g., government officials, technicians, and researchers) and the non-experts (e.g., local community). Table 6-1 presents the components of the assessment methodology concerning the unstructured interviews and the questionnaire. These indicators have been adopted from Alexander et al. (2016) to evaluate flood risk governance in the UK context.

Table 6-1. The assessment methodology components and indicators. The questions are listed in Appendix E.

Section	The number of the questions	Subject	The number of the questions	Indicator
1. Unstructured interviews	24	1. General flood management	17	Knowledge and Awareness, Cooperation and Governance, the Mode of Operation, Financial requirements, Co-benefits
		2. The OA approach	7	
2. Questionnaires	16	1. General flood management issue	8	Knowledge and Awareness, Cooperation and Governance, the Mode of Operation, Financial requirements, Co-benefits, Flexible in Adaptation, Efficiency, and Potential of methodology.
		2. The OA approach issue	8	

In this study a selection of these indicators has been used to assess the perception of the stakeholders on the current flood management practices and the potential benefits of the OA approach in Bangkok. The selection of these eight indicators is based on their relevance for FRM in the context of Bangkok. The selected indicators relate to the subjects: 1) Knowledge and Awareness, 2) Cooperation and Governance, 3) the Mode of Operation, 4) Financial requirements, 5) Co-benefits, 6) Flexibility in Adaptation, 7) Efficiency, and 8) Potential of methodology. A 5-point weighting scale from "**strongly disagree**" to "**strongly agree**" has been used to assess the scoring on these indicators by the interviewees per subject.

Three steps (or rounds) have been followed to complete the assessment. This 3 steps approach has been adapted from Heitz et al. (2009) (see Fig 6-2). The steps involve:

> **Step 1**: Evaluation of the current approach by conducting the interviews and using the questionnaire. The unstructured interviews were focused on general flood related questions aiming to deepen the understanding of the

existing organizational structures and procedures, which come into play
when dealing with floods in Bangkok.

Step 2: *Ex ante* evaluation of the OA approach was conducted using
unstructured interviews, and a questionnaire involving national and local
government-appointed officers and the local community of the case study
area. The objective of this step is to provide insight into the enabling
conditions of the OA approach. The outcomes of step 1 and 2 are presented
in spider diagrams which allow to compare the results between the
stakeholder groups.

Step 3: Assessing the feasibility (enabling conditions) of OA is examined by
comparing the results of step 1 and 2 using the spider diagrams.

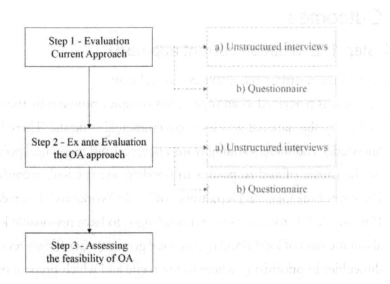

Fig 6-2. Flow chart of the methodology to assess the feasibility of the OA approach

As indicated above the target audience consisted of two groups. The first group
consisted of 31 experts from BMA, members of the national government, and
members of the local government. All of them were on active duty during the flood
disaster which struck Bangkok in 2011. The second group consisted of local
stakeholders from Lab Krabang, the case study area, and involved a group of 61

people (including citizens and local government officials). All interviewees participated in the roundtable discussions. Table 6-2 provides an overview of the origin of the 92 unstructured interviewees.

Table 6-2. Origin and number of interviewees.

Origin (stakeholder groups)	Number of interviewees
1. National government officials	3
2. BMA officials	28
3. Local government officials from Lad Krabang	10
4. Citizens from Lad Krabang	51

6.4 Outcomes

6.4.1 Step 1: Evaluation current approach

The results of the unstructured interviews reveal that:

1. Flooding is perceived as an urgent, but complex problem by the citizens as well as by the national and local governmental officials. 'Therefore, more knowledge and understanding from a practical and theoretical point of view on the causes of and responses to flooding, are needed', according to the Director of the General Department of Public Works and Town & Country Planning. While most citizens acknowledged to have reasonable knowledge about the risks of local flooding, the local government officials reveal to have difficulties in prioritizing where to intervene and which preparatory actions need to be undertaken (in the community areas, village areas, commercial areas, and open spaces). Moreover, the majority of the interviewees has indicated that the selection of flood adaptation measures is not based on proper investigations and lacks community feedback in the decision-making process. According to the interviewees these shortcomings are partially due to the strict top-down nature of the underlying governance process. As a consequence, the effectiveness of measures taken by the local government

are conceived as low. On the other hand, the citizens who have an understanding of the local flood risks, have indicated to lack structured knowledge about how to respond to mitigate flood risks, let alone to have access to tools required to engage in the process of selection and customization of sustainable flood mitigation solutions.

The unstructured interviews with BMA officials reveal that local governments are committed to organize campaigns, trainings and workshops to increase the knowledge and awareness of the community in dealing with floods. However, the information provided by the community indicates that district officials have indeed organized such gatherings, albeit limited to as single meeting shortly after the extreme flooding of 2011.

2. The implementation of the 'Dike' system project along the canal and the 'Monkey Cheek' strategy to mitigate flood risk of Bangkok, which are part of the DDS plan from 2016, are lagging behind. At the time of writing of this chapter, over 80 percent of the flood protection measures are still in progress (personal communication Chief Resilience Officer, 2020). These projects involve the dike system along the Chao Phraya River, various canals, as well as a giant drainage tunnel. Operations which have been carried out by BMA comprises cleaning of the main canal and pipes in order to increase the capacity of the drainage system. Other examples of activities executed by this organization are: collection of garbage and plant wastewater along the canal and survey of the illegal buildings that were built alongside the canal. The majority of the interviewees share the opinion that the national government structure is complex, and the duration of the operations is usually delayed due to bureaucratic processes resulting from centralized decision making. At present, the priority and selection of interventions to be implemented to mitigate the risk of flooding seem to be ad random, and not part of a cohesive long-term flood protection strategy.

3. With respect to cooperation and governance, it was found that various

measures issued by the BMA officials have been successful and well appreciated by the interviewees. These measures include (amongst others) 1) enforcing regulations to removing houses that have been built on and along the canals which is public property, and 2) providing incentives to motivate people to increase the local water storage capacity of their property by allowing an increase of the floor area ratio (FAR) for buildings (the FAR is set by the city planning law).

4. In terms of finance, it was found that the Bangkok budget allocation system came up with an annual budget proposed by the central government. Current urban drainage budgets, which are primarily derived from local taxation, are valued insufficient to finance the required flood mitigation measures.

5. With regards to co-benefits of flood adaptation measures, it was found that the national government sector lacks the (provision of) tools to account for multiple benefits of adaptation measures and in particular, of guidance to identify associated benefits to quantifiable outcomes. At the same time, it is the public sector which is allegedly in the position to identify the benefits which can be obtained from flood prevention efforts.

In summary, governmental officials (i.e. national, BMA, and local) are aware of the existing governance caveats related to the governance structure and responsibilities associated with FRM, which hampers an integrative approach, the operations and coordination of the projects. Finance of flood risk mitigation projects tends to favors large scale interventions (with long lead times) over small, locale scale ones. These affect the planning and interfere with solving future flood events. The community lacks knowledge on how to properly solve flood problems with sustainable measures. Furthermore, people are dependent on local government assistance as they provide the financial support which allows them to act. Most of the people (90 percent according to Leong et al. 2015), has the opinion that actions of solving the flood problem are the responsibility of the national and local government. This creates a conflict between the local government and the citizens, as revealed during the interviews. There seems to be a clear gap between adaptation actions initiated by the local government and citizens. Unclear management at the level of the national government with respect to operations and finance also causes distrust in the community. Therefore, the outcomes of the interviews substantiate the earlier identified conflict between national and local level. Flood protection measures, such as dikes and retention areas, have been built after land expropriation, but the needs of the community and citizens have not been taken into account during this process. Benefits to the country as a whole were mentioned as a motivation for these measures. However, the interviewees also indicated that the real reason behind these decisions were driven by hidden benefits for developers of the private sector.

The results of the questionnaire (see Appendix E) are depicted in Fig 6-3 as a radar chart capturing the 8 indicators. These radar charts are depicting the respondent answers regarding the perceived understanding of the existing organizational structures and procedures for the two groups (national and city institutions (group 1 and 2) and local institutions and citizens (group 3 and 4)).

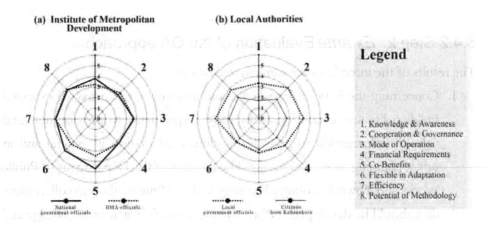

Fig 6-3. Comparison performance of FRM of the two groups (group 1 and 2 and group 3 and 4). Dotted and dashed lines represent the opinion of BMA officials and local government officials, respectively, while the thick and light line represents the views of national government officials and citizens from Kehanakorn village 2, respectively

It follows from Fig 6-3, left diagram that the BMA officials gave a slightly lower score than the national government officials on the indicators, knowledge & awareness, co-benefits and flexible adaptation. The scores on the other indicators are similar. These results indicate that the BMA officials perceive to have a 'reasonable' understanding of the more knowledge intensive subjects, which appears at a slightly lower level than of their peers at the national government.

Fig 6-3, right diagram shows that the citizens gave a (considerably) lower score of most of the indicators compared to the local government. With the exception of the innovative subjects co-benefits and the flexibility of the current FRM. Citizens typically evaluate the current FRM practices lower than the local government.

In conclusion, both the national government and BMA have a very similar perception on the formal aspects of the current FRM practices. The differences in perception between the local government and the citizens on the current FRM practices are much larger: the citizens value their understanding and effectiveness of the current practice as relatively low. This is an indication that the services offered by BMA officials on FRM do not fulfill the expectations of the citizens yet.

6.4.2 Step 2: *Ex ante* Evaluation of the OA approach

The results of the interviews are summarized below:

1. Concerning the issue of knowledge and awareness; the proxy of expected building lifetime is considered by most respondents to provide new and valuable information to consider in climate adaptation in general and in urban planning more specifically. The director of the Department of Public Works and Town & Country Planning said that "the method to collect these data should be developed further". There is clearly a lack of knowledge and tools to guide decision making about where and when to adapt. This proxy could be useful and used by all the stakeholders involved.

2. Concerning the mode of operation; there is a lack of communication and thus coordination between the national government and the local government, planning, and strategic decision making. This shortcoming will complicate the implementation and mainstreaming of the OA approach. Nevertheless, respondents consider the approach as feasible. The BMA should take leadership and initiate and manage implementation of the OA approach, providing that there is a concrete plan (incl. an assessment of the EOLC of buildings in a specific area) as well as an overarching high-level strategy, which provide the top-down boundary conditions for that specific area.

3. In terms of cooperation and governance; enforcing new policies and regulations in FRM is challenging for the current national government. There is a lack of motivation although there is a general sense that the current ones are not adequate particularly in bridging the gap between interventions at national and local level. The OA approach aims to improve this by devolving responsibilities from national to the local level and to develop a common approach with all involved parties. Moreover, this should ease the integration of flood adaptation measures at both large and small scale easier.

4. In terms of the financing requirements; according to the respondents the OA approach has potentials to generate "bankable" projects needed to acquire the required budget from the central government, as, due to the precise operational planning process and multiple value capturing, the risk of malinvestment will be relatively small and the return on investment will be stable and for the long run. This will also support the mobilization of additional funding from various other sources like the local private sector, global private agencies or cloud sourcing.

5. Concerning the co-benefits of the OA approach; most of the respondents stated that the OA process allows all stakeholders to work together (national and local governments as well as the local community). It also provides the opportunity for citizens and local government to discuss and share their concerns regarding the procedures for implementing flood mitigation

measures, with the national government. The OA approach provides a long-term strategic plan, where stakeholders have the opportunity to approve or disapprove agreements.

6.4.3 Step 3: Assessing the feasibility

The results from Step 1 and 2 allow to draw some first conclusions on the feasibility of the OA approach. It has become clear that the OA approach calls for an alignment of adaptation opportunities at different spatial and temporal scales and for having the proper governance levels in place needed to support these alignments. From a top-down perspective, a long-term oriented structured design and planning process is essential. Yet this requires a flexible flood management approach that is tailored to serve the local requirements and expectations. In this context, a few requirements emerge from the results described above. These are:

- Knowledge and Awareness needs to be increased,
- Identifying Adaptation Potential and opportunities,
- Cooperation at different levels,
- Budget distribution into the local area,
- Synchronizing the relationship between practical requirements and policies and regulations.

Table 6-3 provides the key governance challenges and enabling conditions: it compares the major features of the perception of the national government and to the local community and inhabitants, and the enabling conditions of the OA approach extracted from this comparison.

Table 6-3. Key governance challenges and enabling conditions (to address these challenges) at different levels for effective implementation of the OA approach in Bangkok.

Characteristic	Challenge		Enabling conditions
Level	National Government	Local inhabitant	
Flood management	• Looking only at extreme flood events. • Impact on regional to national level, caused by floodwater coming from upstream and/or elsewhere	• Focus on local (moderate & frequent) flood events. • Local impacts caused by local flooding	• Acknowledging the importance of the interrelationship between extreme (rare) and local (moderate and frequent) flood events.
Policy	• The government relies on reactive policies as a way to mitigate flood damage. The focus is on plans with a short-term scope. • Long-term strategies and adaptive plans are not yet being considered	• Increased FAR of the property when a water storage solution was applied. • Focus on preparedness and short-term strategies such as water pumping machines.	• Short-term actions should be aligned with a long-term (adaptive) strategy.
Lead time of projects	• Long (years to decade) and uncertain due to lengthy and bureaucratic procedures	• Short (<yr.) but uncertain, incentives easily fade away, momentum has to be seized	• Flexible, short-time procedures
Knowledge and awareness	• Capacity building and awareness raising in FRM have not received priority yet for the government.	• Need to learn and share knowledge and technology for dealing with local flood events are low • Lack of awareness	• Embracing the importance of collective learning, knowledge sharing and awareness raising are key to foster decision-makers and citizens.

Cooperation and governance	• Reliance on formal procedures, regulations and laws to water management	• Lack of time and means to participate in collective (multi-stakeholders) processes • Lack of shared platforms to ease communication	• New channel or platform to engage with stakeholders to align interest and co-create
Financial requirements	• Dedicated budgets for mono-functional infrastructure. • Large investments	• Using own often fragmented funding • Small budgets	• Budget clustering to leverage with other funding sources, reduced cost of implementing flood adaptation measures by maximizing co-benefits and alignment long-term strategy
Multiple benefits	• Emphasis on engineering efficiency and optimality in flood protection resulting in larger (underground) drainage systems.	• Single focus on flood mitigation inside the asset space	• Multiple functions are essential features of measures: more green space, open spaces, and recreation areas

6.5 Conclusions and recommendations

Awareness is growing that FRM calls for collaborative action involving stakeholders across governmental levels and local communities. OA is conceived as an approach which fosters and guides such collaboration and expands the options for intervention. This study indicates that OA is recognized by the stakeholders involved as an innovative approach to be used in urban planning, designed to assist them to share a common vision on what/when/where to adapt in the future, taking also into account uncertain conditions like, e.g., climate change. For example, it provides members of the community (financial and social) incentives to take action by themselves in a manner which is aligned with the local and/or national FRM policy. For the local government (e.g., the district office), OA legitimates to engage with the local community in order to make flood adaptation measures more effective, by having the local community properly involved. Finally, it provides more options for intervening with flood adaptation measures, rather than controlling and commanding the people to follow a strategy decided through top-down governance. This approach creates a shared space for stakeholders, which can be used to organize cooperation between them by allowing them to work on a common solution to the underlying problem.

The OA approach is about integrating flood adaptation into urban renewal cycles. This inherently means that it is a *design-driven* process and therefore requires designers to be central in the process and in the OA team. In Thailand, the designers may sit within BMA or come from external sources. For example: social experts, ecologists, road and infrastructure experts, engineers, and especially architects. In other countries this might vary, depending on the level of engagement of government agencies in urban renewal either as initiating actor and facilitator or as a mere legislator with limited active involvement in the restructuring of urban areas.

While OA depends on a lifecycle perspective, few traditional parties in the urban renewal process are involved in monitoring and managing the lifecycle of buildings. This is in contrast to public space and infrastructure which, depending on the context, are often actively maintained and refurbished by government agencies. Extending this role to especially the housing sector would require homeowners to adopt a more rational approach to depreciate their properties. This might require a shift in roles and responsibilities that to some extent might be adopted by the designing community: architects and planners.

In building design, architects and designers generally do not acknowledge food risk as a design consideration. Typically, only when constrains are expressed .in a particular building code or regulation, those constrains are incorporated into a design very much like a fire regulation or zoning law. This is partly due to the fact that they have to follow a fixed program or brief which describe the requirements that need to be reconciled and accommodated. Hence, the role of an architect usually comes down to the design of a building (e.g., housing and apartment), and a rearrangement of the area. This study suggests that the OA approach offers designers to take a new role in the design process as it offers an opportunity for them to take a leadership role, to take a pro-active, process orientated role, by bringing national and local government officials, engineers, and citizens living together in the affected areas with the ultimate aims to maximize the benefits for each stakeholder.

Furthermore, providing and sharing knowledge, awareness, and experience in terms of climate change and flood prevention to stakeholders and officers is significant. This could be done, for example, in terms of the procedure level by using media to educate people, and at the policy level by using zoning as a governance for new building constructions. Additionally, increasing the channel of support for stakeholders will improve data collection methods as well as the implementation of measures.

This study indicates that OA also provides a governance framework to integrate the **'top-down'** and **'bottom-up'** approach in FRM. In Thailand, which serves as the context for the case-study, both approaches are disconnected trajectories in the current practice in Bangkok operating at different scales. The clustered approach calls for an alignment between the governmental institutions. A better understanding and demonstration are needed how OA can generate additional value in practice.

Mainstreaming of the OA approach in Thailand will require a transition flood management practices and governance, which will likely not happen overnight. To start this transition a first step is to perform a LCA of the buildings and other prime assets in the various districts of Bangkok. This data will likely drive institutions such as BMA, but also local stakeholders (incl. the private sector) to identify 'low hanging fruit', obvious opportunities for OA on the short-term to cluster and collectively adapt urban neighborhoods. This could be in the form of some pilot projects. These have to be initialized to build confidence in the feasibility and practicalities of this new approach. The establishment of these pilots will also allow to test innovative business models which would enable to identify bankable arrangements underlying these OA projects and provide pathways for wider uptake.

CHAPTER 7
Conclusions and recommendations

7.1 Introduction

Bangkok experiences an increase in flood losses. Due to population growth, the city is expanding rapidly. Without adaptation measures, and assuming a four-degree Celsius temperature increase, the city is expected to experience around 40 percent inundation of the total city area by an extreme rainfall event every year and 15 centimeters (cm) sea-level rise (SLR) by 2030 (WB, 2019). Adaptation to flood risk is happening, but up till now an effective city-wide FRM strategy is still lacking. There is a need for a more pro-active and strategically planned approach to FRM which leads to better coordination at the local/grass root level. At the same time, Bangkok is a highly dynamic city which changes over time as buildings and infrastructure are continuously renewed. These renewal activities provide opportunities to contribute to a more devolved FRM requiring coordination across scales and sharing of responsibilities across stakeholders.

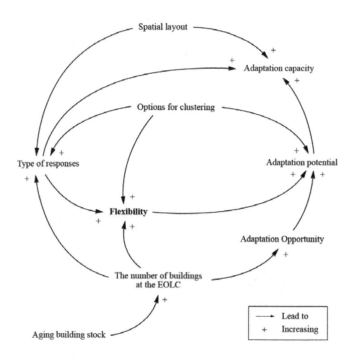

Fig 7-1. Causal diagram capturing the OA process underlying this dissertation

The research described in this dissertation aims to contribute to a better understanding and use of Opportunistic Adaptation (OA), an approach to assists cities to climate change. The underlying processes comprise both positive and negative feedback loops connecting variables of which the *Adaptation Potential* (AP) is identified as one of the key variables (see Fig 7-1). Hence, the first objective of this dissertation is to assess the AP derived from opportunities to intervene in the city. In this dissertation, the AP is defined as the number of opportunities to adapt buildings and infrastructure in the future arising from renewal. Knowing the AP and subsequently the adaptation opportunities are conditional to develop a FRM strategy. The second objective is about how to use the AP to develop a flexible and inclusive FRM strategy. The flexibility of the strategy is determined by the number of adaptation opportunities available *over a period of time* as well as the number of adaptation options available *at a certain time interval.* An urban district of Bangkok, Lad Krabang, has been selected as a case study area. An additional objective of this research is to gain insight into when and how to select flood adaptation measures. This objective relates to the technical and governance aspects of the OA approach. With respect to the latter, specific attention has been paid to the organizational and procedural conditions required for wider uptake and mainstreaming of this new approach.

7.2 Conclusions

Cities are adapting to a changing flood risk. In many cities, this process is largely autonomous and not planned. The existing urban fabric of cities changes over time as buildings and infrastructure are continuously being maintained, renovated and replaced. These dynamics provide opportunities to adapt the city to changing hydrological conditions and to reduce flood risk. The current practice of urban FRM does not fully exploit or utilize this *AP*. In order to do so, a detailed understanding of the opportunities for timely intervening in the urban fabric is required. The adaptive potential is closely linked to the lifecycle of buildings. Most cities have a heterogeneous building stock resulting from a fragmented development history. Yet, a large number of assets are continually changing over time. In order to effectively use the AP of cities, an opportunistic planning and transformation strategy with a long-time horizon is being proposed in this research. This strategy should be inclusive in order to identify and appraise multiple benefits resulting from adaptation measures. This in turns calls for supporting tools to engage stakeholders and to collaboratively arrive at an agreement on when and where to adapt taking the various spatial scales into consideration (and having a long-term time horizon). This study has shown that one of the benefits of the newly developed OA approach is that it allows to identify clusters of buildings and adjacent areas which can be adapted at the same time in an integrated, combined manner. Clustering expands the number of adaptation options as it also considers options beyond the building level and magnifies the functioning of green infrastructure as more green space will become connected and available for ecological and recreational services. Adaptation options are identified, selected and integrated into the design of the newly built or to be retrofitted assets or groups of assets and are implemented when the existing urban assets reach the EOLC (urban renewal cycles). By using a computer simulation model, the outcomes are presented as a decision-tree diagram. This diagram expresses the AR by depicting all possible adaptation options available at different points in time. A scoring has been developed to measure the flexibility of the selected adaptation measures of the decision-tree. This scoring can be computed for single assets or clusters of assets with varying complexity.

Conclusions regarding the sub-questions

1. What are the components of an operational method to identify adaptation opportunities based on autonomous urban renewal cycles?

Adaptation measures are conceived as interventions taken at individual building level, neighborhood level and/or landscape level, which improve the ability of the city to better cope with rainfall and future changes in rainfall. This is operationalized by adopting a lifecycle approach of urban assets, where adaptation measures are installed when assets reach the end of their functional life: their EOLC. Depending on the spatial proximity and differences in ELOC, adaptation measures can be clustered around multiple assets which are depreciated at the same point in time. (e.g., two neighboring dwellings). Based on a review of the literature, this study has identified the following components to be taken into account to assess the EOLC of buildings in a given area:

1. *Asset location.* An inventory of all asset locations for a given area-of-interest is required. Typically, this is provided in the form of a GIS-based map.

2. *Asset construction age.* To estimate the EOLC, the year of construction for each asset is required. This information is often provided in combination with the asset location in a GIS database or it is available in cadastral records. In case this information is not available, it needs to be estimated on the basis of a comparison of similar assets which can serve as a reference and of which this information is available. Alternatively, experts need to be consulted to obtain information about the age of assets.

3. *Lifespan of the assets.* To estimate the future EOLC, the lifespan of the assets needs to be approximated. Figures are typically based on statistical evidence from large scale surveys and derived from time-series data from urban renewal records. To make this operational, the estimated lifespan needs to be associated with specific asset characteristics.

4. *Clustering method.* In order to synchronize the EOLCs of assets to scale up adaptation opportunities, a spatiotemporal clustering method needs to be applied that identifies clusters based on small differences in the estimated EOLCs of assets located in close proximity to each other. These are grouped into distinct sets that can subsequently be treated as a group of assets (e.g., buildings, (green) infrastructure, and public spaces) that can be adapted simultaneously and in an integrated fashion at a common future

point in time.

The clustering characteristics are defined by a set of parameters which expresses the thresholds with respect to maximum differences in EOCLs and in distance between assets on which clusters are composed. This means that both a spatial and temporal tolerance needs to be defined, which is an arbitrary value based on a chosen depreciation scheme. In turn, this depreciation scheme is based on the level of integration of the redevelopment strategy and its organization. If an area is to be redeveloped as a single entity, the area effectively has to be captured in a single cluster. This can be seen as a top-down strategy typically initiated by the government. Obviously, this normalizes large differences in the estimated EOLC. If redevelopment is foreseen at an individual asset-level, a bottom-up strategy needs to be applied. When optimizing the two by maximizing the opportunities from these spatiotemporal relationships, a clustered adaptation strategy is applied.

Flood maps have been used to identify the flood exposure of the case study area, the Lad Krabang district. This has resulted in the identification of "hot spots", areas which exhibit a relatively high flood exposure (based on flood frequency and inundation depth).

2. Which factors influence the Adaptation Rate and capacity and how do they relate to the spatial dimension of a city?

Urban renewal is the key driver of change and thus the dominant factor which controls the AR. Roads and other public urban assets have typically relatively short lifecycles compared to buildings. However, in practice they are often redeveloped at the same time as those of large-scale upgrading programs when adjacent buildings are being renovated or replaced. The AR is a numerical value that defines the rate at which buildings in a particular area-of-interest reach their EOLC within a given period of time.

An understanding of the AR is conditional for the development and implementation of an urban adaptation strategy. It provides information which is required to integrate adaptation planning with other sectoral agendas, such as energy, transport and climate change. The AR assessment process aims to establish a relationship between the actual space, and the temporal characteristics of the assets occupying this space. Clusters of individual buildings can be composed based upon to their EOLC. This study has developed and applied a clustering method which is based on two factors, namely EOLC and distance between the assets. Both factors have been quantified by the terms EOLC-Gaps and search radii.

3. How does Opportunistic Adaptation affect the flexibility of adaptation options over space and time?

Changing conditions often require to adjust or replace adaptation measures. This calls for a certain degree of flexibility. The flexibility is reflected in the choice and design of the adaptation measures (e.g., a measure can be adjusted over time) and the overarching strategy (e.g., the rate of implementation of measures can be decreased or increased). In this study, sequencing adaptation measures (adaptation pathways), has been used to define an adaptation strategy. An adaptation strategy needs to be flexible (speeding up or slowing down the implementation process of adaptation measures or changing the sequence (pathway) of adaptation measures) in order to deal with unforeseen changing conditions. Theoretically the number of adaptation options is decreasing over time. However, this study reveals that in practice, this is not necessarily the case as the number of adaptation options is also dependent on the size of the clustering area, the characteristic of the urban context, and the flood features such as inundation depth. For example, the number of options increases with an increase of the spatial scale considered in the OA approach. The lifespan of a measure has a direct positive correlation with its flexibility. The lifespan of a measure also impacts the flexibility of the adaptation strategy and relates to what is referred to as "early and late adaptation". Adopting adaptation measures that have a long lifespan in an early stage of the transformation process ("early adaptation"), reduces the flexibility of the strategy, as the measure creates a lock-in for future change.

4. What institutional arrangements does Opportunistic Adaptation require to flourish given the different stakeholders involved in urban climate adaptation?

There is a need to better coordinate and align autonomous adaptation (bottom-up initiatives) with planned FRM interventions imposed by the city government and to accelerate the development and implementation of FRM at city governmental level. The workshops and interviews with stakeholders (with representatives ranging from the national and city governmental (BMA) level to the local community level) revealed that the proposed OA approach provided social, economic and procedural (from a governance perspective) incentives to embrace and participate in a collective process of developing a climate adaptation strategy. There is a willingness to take responsibility at all levels and a notion that action is needed. The participants of the workshops expect that the OA approach will foster joint vision making, networking and the development of group skills needed to take a concerted and guided action. More specifically for local communities fund raising opportunities have been identified as a positive spin-off. From the perspective of the designers, it will enable them to take a facilitating role overseeing and coordinating the process rather than contributing to parts of the process (such as delivery the preliminary design). From the point of view of knowledge providers (engineers, architects and planners), it requires that they have access to novel information to apply and exploit the OA approach, such as information on Adaptation Rates, potential adaptation measures and dynamic design principles. In practice, the OA approach gives information about when and where the process should start as the decision-tree captures the information about which measure(s) need to considered on the short term and which could be applied on the longer term. The resulting adaptation pathways, as shown in the workshops, position the stakeholders dialogue in a longer term perspective (beyond the business as usual practice) and allow for a discussion with less defensive behaviour required to develop a common sense of purpose based on a shared vision. This will finally help in arriving at an agreement on a common approach with the stakeholders, on a case by case basis. The workshops conclusions support the premise that the OA approach creates conditions which might decrease social conflict between the government authorities and the people from the local community. Therefor it has potential to create a common language and understanding among the various stakeholders, to work across governance levels, disciplines, and time frames.

Based on the interviews and the workshops with stakeholders it was found that there is an information gap (i.e. knowledge and know-how) stemming from the fact that the availability and sharing of data (e.g., related to flood extend and levels during a flood) to support the current FRM are hampered. Many people know that their property is at risk of being flooded, but they lack the capacity to act. Particularly, information on a long-term FRM strategy of the city government is absent. As a result, they have no perspectives to act on the notion that their flood risk is increasing and that climate change is eroding the level of flood protection. City governments are too much focused on traditional approaches with an emphasis on reducing the probability of flooding using large-scale infrastructure systems such as drainage systems, flood defenses and waterways. They often are not aware of the opportunities available at the local scale, and thus lack the insights to unlock the complementarity of bottom-up approaches to hydraulic engineering approaches. This also holds true for the city of Bangkok, as evidenced by this study and amplified during the workshops. Here, the OA approach has potentials to make a difference by aligning the prevailing modus operandi of top-down governance and uncoordinated bottom-up initiatives as it assists all stakeholders to jointly (i) set aspirations which go beyond flood safety and (ii) explore the consequences of various interventions/measures.

Lead time and associated investment issues of adaptation measures also need to be considered as they have ramifications for the governance of OA. For Bangkok, studies show that upgrading large-scale flood protection infrastructure will reduce the flood risk. However, it requires huge investments and long lead times and comes with residual risk. Bottom-up, small scale interventions have a much shorter lead time and will therefore be needed to sustain and increase the level of flood resilience on the short term and thus to bridge this time interval gap of implementation of large schemes. Hence, from a perspective of finance and lead time of measures an approach in which both top-down and bottom-up interventions are being jointly considered, such as in the OA approach, may overcome these constraints arising from large scale interventions.

In Bangkok, top-down initiatives are typically driven by flood disasters resulting in a master planning approach where experts are playing an important role to consult public organizations such as the national government (ONWR), city governments (BMA), as well as the local government, representatives of NHA, DDS, and the local administration office. The government of the city, acting as leader of the flood risk strategy of Bangkok in 2013 (Dhakal and Shrestha, 2016) is now gradually starting to understand that collaborative and participative citizen approaches should be incorporated in their predominantly reactively driven FRM strategy (Singkran, 2017). Bottom-up initiatives in Bangkok involve citizens from local communities as well as entities from the private sector (e.g., private companies and developers) (Ajibade and Egge, 2019). However, interviews with representatives from the national and local governments pointed out that convincing the local community to engage in a new method as part of the OA procedure and to take responsibility is very challenging, due to the fact that knowledge and awareness of the financial ramifications and time consuming procedures are missing.

7.3 Reflection and recommendations

7.3.1 Recommendations for further development of the OA approach

Based on the research findings from this dissertation, the following recommendations for future research are given:

1. In this study, the OA approach has exclusively focused on buildings and renewal of buildings after reaching the end of the lifetime. The challenge for future studies will be to expand the scope to other elements of the urban fabric, such as roads, drainage systems, and public spaces. Moreover, when a building reaches its EOLs, it is assumed in this study that the building will be replaced completely. However, other (intermediate) interventions should be considered as well, such as renovation and maintenance. These interventions may also provide opportunities for adaptation.

2. The second recommendation is to consider changing the purpose (function) of buildings/areas as an adaptation strategy. That is because some functional requirements are more flood vulnerable than others: changing the function of a building is a strategy to adapt to (climate) change. However, it is difficult to predict when and for which purpose a building can actually change. A typical example of functional change in this context is when a flood vulnerable residential function at the ground floor of a building changes into a less flood vulnerable industrial or commercial function. Changing the function of a building does not necessarily depend on the building lifespan alone. Also, other factors such as financial or regulatory ones can be drivers for function change. Finally though, including functional changes as an adaptation option will most likely create such an degree of freedom in that a structured approach to assessing future adaptation options becomes intractable.

3. The third recommendation relates to the question on how to strike a balance between increasing the scale (clustering sizes) and decreasing the complexity (urban elements)? Complexity is related to urban elements, such as roads or

public spaces, which need to be taken into account in the planning process. The OA approach offers a general guide to different types of clustering. The clustering strategy needs to be considered when dealing with a high degree of complexity with respect to the existing urban elements. The best possible integration method needs to be identified, in order to answer the question of what the optimal size of a cluster should be. The set of tools used in this process should be able to assess and provide more insight into research conducted on the clustering radius by using different urban typologies such as center and suburbs. These values should not simply be decided by the stakeholders or be established only by case studies. Instead, further inquiries into the future impact and an overview of all involved factors are required to achieve the best results.

4. Further research will be needed to assess the costs and benefits of the proposed adaptation interventions and their effectiveness, which are beyond the scope of the dissertation. This will require additional, detailed flood risk modelling of the case study area for the current situation (present time) and future projections (using various scenarios).

7.3.2 Recommendations for applying OA approach in the Lad Krabang district

1. The OA approach has been presented to the stakeholders including local inhabitants, focusing on the workflow required by the methodology and on the difficulties encountered when applying the procedure (CHAPTER 6). The biggest challenge, however, is about how to effectively communicate the benefits of the OA approach? This aspect needs further attention and in particular the role which dynamic designs may have.

2. Based on observations gathered from the interviews on the case study, it was found that focusing on certain privileged groups, the outcomes may be detrimental for the wider community. Promoting the OA approach should not be exclusively driven by local interests, but instead from an urban

regulatory perspective, it should strike a balance between local, regional and national interests. This will diversify and expand the number of stakeholders to be motivated to get involved and even to be part of the design process.

3. The OA approach focuses on flood risk at the local level- ranging from individual buildings to neighborhood level. It provides options to the local community and authorities to act. To effectively implement the OA approach, it needs to be supported by economic tools to assess the benefits and costs of the various adaptation measures (and pathways) that have been identified. The economic tools are a crucial component here. However, it goes beyond the scope of this research.

4. The OA approach is based on making certain decisions, at certain points in time, which are reflected in the decision-tree diagram. A designer has the role to simplify the decision process as much as possible. This will require new skills and knowledge, which need to be developed through dedicated training programs.

5. Adaptation opportunities are not only affected by the lifecycle of the buildings, but also by the lifecycle of the tenancy contract (land or building), which is owned by the Crown Property Bureau (CPB) and the national government (Ouyyanont, 2008; Pimonsathean, 2012). This needs to be taken into consideration and require further study.

References

Aerts, J.C.J.H., Botzen, W.J.W., Emanuel, K., Lin, N., de Moel, H. and Michel-Kerjan, E.O. 2014. Climate adaptation. Evaluating flood resilience strategies for coastal megacities. *Science* 344(6183), pp. 473–475.

Adelekan, I.O., 2010. Vulnerability of poor urban coastal communities to flooding in Lagos, Nigeria. *Environment and urbanization, 22*(2), pp.433-450.

Ajibade, I., & Egge, M. 2019. Synergies and opportunities for transformation. *Achieving the Sustainable Development Goals: Global Governance Challenges.*

Alexander, M., Priest, S., & Mees, H. 2016. A framework for evaluating flood risk governance. *Environmental Science & Policy, 64*, 38-47.

Alhassan, S., & Wade, L. H. 2017. Challenges and opportunities for mainstreaming climate change adaptation into WASH development planning in Ghana. *International journal of environmental research and public health*, 14, no. 7, 749.

Anderson, H.C. 2014. Amphibious architecture: Living with a rising bay.

Anon 2017. *Flexibility in adaptation planning: when, where and how to include flexibility for increasing urban flood resilience.* CRC Press.

Anvarifar, F., Zevenbergen, C., Thissen, W. and Islam, T. 2016. Understanding flexibility for multifunctional flood defences: a conceptual framework. *Journal of Water and Climate Change* 7(3), pp. 467–484.

Araos, M., Ford, J., Berrang-Ford, L., Biesbroek, R. and Moser, S. 2016. Climate change adaptation planning for Global South megacities: the case of Dhaka. *Journal of Environmental Policy & Planning* 19(6), pp. 1–15.

Ashley, R., Garvin, S., Pasche, E., Vassilopoulos, A., & Zevenbergen, C. 2007. *Advances in urban flood management.* CRC Press.

Ashley, R., Lundy, L., Ward, S., Shaffer, P., Walker, L., Morgan, C., Saul, A., Wong, T. and Moore, S. 2013. Water-sensitive urban design: opportunities for the UK. *Proceedings of the Institution of Civil Engineers - Municipal Engineer* 166(2), pp. 65–76.

Ayers, J. M., Huq, S., Faisal, A. M., & Hussain, S. T. 2014. Mainstreaming climate change adaptation into development: a case study of Bangladesh. *Wiley Interdisciplinary Reviews: Climate Change, 5*(1), 37-51.

Ayog, J. L., Tongkul, F., Mirasa, A. K., Roslee, R., & Dullah, S. 2017. Flood risk assessment on selected critical infrastructure in Kota Marudu Town, Sabah, Malaysia. In *MATEC Web of Conferences* (Vol. 103, p. 04019). EDP Sciences.

Batica, J., & Gourbesville, P. 2016. Resilience in flood risk management—a new communication tool. *Procedia Engineering, 154*, 811-817.

Bentley, J. L., Stanat, D. F., & Williams Jr, E. H. 1977. The complexity of finding fixed-radius near neighbors. *Information processing letters, 6*(6), 209-212.

Botzen, W. J., & Van Den Bergh, J. C. 2008. Insurance against climate change and flooding in the Netherlands: present, future, and comparison with other countries. *Risk Analysis: An International Journal, 28*(2), 413-426.

Bretz, H. 1998. Challenges in tunnel engineering in Southeast Asia. *Tunnel Construction*, 73-82.

Bloemen, P., Reeder, T., Zevenbergen, C., Rijke, J. and Kingsborough, A. 2017. Lessons learned from applying adaptation pathways in flood risk management and challenges for the further development of this approach. *Mitigation and Adaptation Strategies for Global Change* 23(7), pp. 1–26.

Bradford, R. A., O'Sullivan, J. J., Van der Craats, I. M., Krywkow, J., Rotko, P., Aaltonen, J., & Mariani, S. 2012. Risk perception--issues for flood management in Europe. *Natural Hazards & Earth System Sciences, 12(7)*.

Buurman, J. and Babovic, V. 2016. Adaptation Pathways and Real Options Analysis: An approach to deep uncertainty in climate change adaptation policies. *Policy and Society* 35(2), pp. 137–150.

Byg, A. and Herslund, L. 2016. Socio-economic changes, social capital and implications for climate change in a changing rural Nepal. *GeoJournal* 81(2), pp. 169–184.

Campillo, G., Mullan, M. and Vallejo, L. 2017. Climate Change Adaptation and Financial Protection.

Camporeale, P. 2013. Genetic Algorithms Applied to Urban Growth Optimization.

Cao, H., Liu, J., Chen, J., Gao, J., Wang, G. and Zhang, W. 2019. Spatiotemporal patterns of urban land use change in typical cities in the greater mekong subregion (GMS). *Remote sensing* 11(7), p. 801.

Cao, L., Zhang, Y., Lu, H., Yuan, J., Zhu, Y., & Liang, Y. 2015. Grass hedge effects on controlling soil loss from concentrated flow: A case study in the red soil region of China. *Soil and Tillage Research, 148*, 97-105.

Cardno, C. A. 2014. Rebuilding by Design. *Civil Engineering Magazine Archive, 84(7)*, 66-80.

Chu, E., Anguelovski, I., & Roberts, D. 2017. Climate adaptation as strategic urbanism: Assessing opportunities and uncertainties for equity and inclusive development in cities. *Cities, 60*, 378-387.

Clemens, M., Rijke, J., Pathirana, A., Evers, J., & Hong Quan, N. 2016. Social learning for adaptation to climate change in developing countries: Insights from Vietnam. *Journal of Water and Climate Change, 7(2)*, 365-378.

Confederation, S. 2017. DISASTER RISK FINANCE COUNTRY NOTE: SERBIA. *World*.

Cotterill, Sarah, and Louise J. Bracken. "Assessing the Effectiveness of Sustainable Drainage Systems (SuDS): Interventions, Impacts and Challenges." Water 12, no. 11 (2020): 3160.

Darajeh, N., Idris, A., Masoumi, H. R. F., Nourani, A., Truong, P., & Sairi, N. A. 2016. Modeling BOD and COD removal from Palm Oil Mill Secondary Effluent in floating wetland by Chrysopogon zizanioides (L.) using response surface methodology. *Journal of environmental management, 181*, 343-352.

Dhakal, S., & Shrestha, A. 2016. Bangkok, Thailand. *Cities on a Finite Planet: Towards Transformative Responses to Climate Change. Routledge, UK*, 63.

Dolman, N., & Ogunyoye, F. 2018, April. How water challenges can shape tomorrow's cities. In *Proceedings of the Institution of Civil Engineers-Civil Engineering* (Vol. 171, No. 6, pp. 22-30). Thomas Telford Ltd.

Douglass, M. 2016. The urban transition of disaster governance in asia. In: Miller, M. A. and Douglass, M. eds. *Disaster governance in urbanising asia.* Singapore: Springer Singapore, pp. 13–43.

Duijn, M. and van Buuren, A., 2017. The absence of institutional entrepreneurship in climate adaptation policy–in search of local adaptation strategies for Rotterdam's unembanked areas. Policy and Society, 36(4), pp.575-594.

Dutta, D. 2011. An integrated tool for assessment of flood vulnerability of coastal cities to sea-level rise and potential socio-economic impacts: a case study in Bangkok, Thailand. *Hydrological sciences journal, 56*(5), 805-823.

Eichler, H.G., Baird, L.G., Barker, R., Bloechl-Daum, B., Børlum-Kristensen, F., Brown, J., Chua, R., Del Signore, S., Dugan, U., Ferguson, J., Garner, S., Goettsch, W., Haigh, J., Honig, P., Hoos, A., Huckle, P., Kondo, T., Le Cam, Y., Leufkens, H., Lim, R. and Hirsch, G. 2015. From adaptive licensing to adaptive pathways: delivering a flexible life-span approach to bring new drugs to patients. *Clinical Pharmacology and Therapeutics* 97(3), pp. 234–246.

Escarameia, M., Walliman, N., Zevenbergen, C., & de Graaf, R. 2015, June. Methods of assessing flood resilience of critical buildings. In *Proceedings of the Institution of Civil Engineers-Water Management* (Vol. 169, No. 2, pp. 57-64). Thomas Telford Ltd.

Evans, G. 2015. Design for hydrocitizens: architectural responses to the defend-retreat-attack scenario. *Sustainable Mediterranean Construction.*

Field, C. B., Barros, V., Stocker, T. F., & Dahe, Q. (Eds.). 2012. *Managing the risks of extreme events and disasters to advance climate change adaptation: special report of the intergovernmental panel on climate change.* Cambridge University Press.

Field, C.B. 2014. *Climate change 2014–Impacts, adaptation and vulnerability: Regional aspects.* Books.google.com.

Francesch-Huidobro, M., Dabrowski, M., Tai, Y., Chan, F. and Stead, D. 2017. Governance challenges of flood-prone delta cities: Integrating flood risk management and climate change in spatial planning. *Progress in planning* 114, pp. 1–27.

Freund, Y., & Mason, L. 1999, June. The alternating decision tree learning algorithm. In *icml* (Vol. 99, pp. 124-133).

Friend, R. and Thinphanga, P. 2018. Urban transformations across borders: the interwoven influence of regionalisation, urbanisation and climate change in the mekong region. In: Miller, M. A., Douglass, M., and Garschagen, M. eds. *Crossing Borders.* Singapore: Springer Singapore, pp. 97–116.

Fujimoto, M., Puangchit, L., Sugawara, F., Sripraram, D., Jiamjeerakul, W., & Kato, H. 2016. Carbon Sequestration Estimation of Urban Trees in Parks and Streets of Bangkok Metropolitan, Thailand. *Thai J. For, 35*(3), 30-41.

Gale, E. L., & Saunders, M. A. 2013. The 2011 Thailand flood: climate causes and return periods. *Weather, 68*(9), 233-237.

Gersonius, B. 2012. *The resilience approach to climate adaptation applied for flood risk.* Repository.tudelft.nl.

Gersonius, B., Nasruddin, F., Ashley, R., Jeuken, A., Pathirana, A., & Zevenbergen, C. 2012. Developing the evidence base for mainstreaming adaptation of stormwater systems to climate change. *Water research, 46*(20), 6824-6835.

Gersonius, B., Ashley, R., Pathirana, A. and Zevenbergen, C. 2013. Climate change uncertainty: building flexibility into water and flood risk infrastructure. *Climatic Change* 116(2), pp. 411–423.

Gersonius, B., Ashley, R., Jeuken, A., Pathinara, A. and Zevenbergen, C. 2015. Accounting for uncertainty and flexibility in flood risk management: comparing Real-In-Options optimisation and Adaptation Tipping Points. *Journal of Flood Risk Management* 8(2), pp. 135–144.

Gersonius, B., Rijke, J., Ashley, R., Bloemen, P., Kelder, E. and Zevenbergen, C. 2016. Adaptive Delta Management for flood risk and resilience in Dordrecht, The Netherlands. *Natural Hazards* 82(S2), pp. 201–216.

Gnansounou, E., Alves, C. M., & Raman, J. K. 2017. Multiple applications of vetiver grass–a review. *International Journal of Environmental Sciences, 2*, 125-141.

Gu, P., Hashemian, M., & Nee, A. Y. C. 2004. Adaptable design. *CIRP Annals-Manufacturing Technology, 53*(2), 539-557.

Guha-Sapir, D., Vos, F., Below, R. and Ponserre, S. 2012. Annual disaster statistical review 2011: the numbers and trends.

Haasnoot, M., Middelkoop, H., Offermans, A., Beek, E. van and Deursen, W.P.A. van 2012. Exploring pathways for sustainable water management in river deltas in a changing environment. *Climatic Change* 115(3–4), pp. 795–819.

Haasnoot, M., Kwakkel, J.H., Walker, W.E. and ter Maat, J. 2013. Dynamic adaptive policy pathways: A method for crafting robust decisions for a deeply uncertain world. *Global Environmental Change* 23(2), pp. 485–498.

Hallegatte, S., Green, C., Nicholls, R. J., & Corfee-Morlot, J. 2013. Future flood losses in major coastal cities. *Nature climate change, 3*(9), 802.

Hara, Y., Takeuchi, K. and Okubo, S. 2005. Urbanization linked with past agricultural landuse patterns in the urban fringe of a deltaic Asian mega-city: a case study in Bangkok. *Landscape and Urban Planning* 73(1), pp. 16–28.

Hoornweg, D., Freire, M., Lee, M.J., Bhada-Tata, P. and Yuen, B. 2011. *Cities and climate change: responding to an urgent agenda.* The World Bank.

Huong, H.T.L. and Pathirana, A. 2013. Urbanization and climate change impacts on future urban flooding in Can Tho city, Vietnam. *Hydrology and Earth System Sciences* 17(1), pp. 379–394.

Huq, S., & Reid, H. 2004. Mainstreaming adaptation in development.

Jha, A.K., Bloch, R. and Lamond, J. 2012. Cities and flooding: a guide to integrated urban flood risk management for the 21st century. The World Bank.

Jha, A.K., Miner, T.W. and Stanton-Geddes, Z. 2013. *Building urban resilience: principles, tools, and practice.* Jha, A. K., Miner, T. W., and Stanton-Geddes, Z. eds. The World Bank.

Johannessen, Å., Rosemarin, A., Thomalla, F., Swartling, Å. G., Stenström, T. A., & Vulturius, G. 2014. Strategies for building resilience to hazards in water, sanitation and hygiene (WASH) systems: the role of public private partnerships. *International Journal of Disaster Risk Reduction, 10*, 102-115.

Jones, S., Tefe, M. and Appiah-Opoku, S., 2015. Incorporating stakeholder input into transport project selection–A step towards urban prosperity in developing countries?. *Habitat International, 45*, pp.20-28.

Jonkman, S. N., & Dawson, R. J. 2012. Issues and Challenges in Flood Risk Management—Editorial for the Special Issue on Flood Risk Management.

Karimpour, M., Belusko, M., Xing, K., & Bruno, F. 2014. Minimizing the life cycle energy of buildings: Review and analysis. *Building and Environment, 73*, 106-114.

Keokhumcheng, Y., & Tingsanchali, T. 2012. Flood Hazard Assessment of the Eastern Region of Bangkok Floodplain, Thailand. In *2nd International Conference on Biotechnology and Environment Management.*

Kittipongvises, S., & Mino, T. 2015. Perception and communication of flood risk: lessons learned about Thailand's flood crisis of 2011. *Applied Environmental Research, 37*(1), 57-70.

Kokx, J.M.C. and Spit, T.J.M., 2012. Increasing the adaptive capacity in unembanked neighborhoods? An exploration into stakeholder support for adaptive measures in Rotterdam, the Netherlands. American Journal of Climate Change, 1, pp.181-193.

Koop, S.H.A. and van Leeuwen, C.J. 2015. Assessment of the sustainability of water resources management: A critical review of the city blueprint approach. *Water Resources Management* 29(15), pp. 5649–5670.

Kraslawski, A. and Turunen, I., 2013. *23rd European Symposium on Computer Aided Process Engineering.* Elsevier.

Laeni, N., van den Brink, M. and Arts, J. 2019. Is Bangkok becoming more resilient to flooding? A framing analysis of Bangkok's flood resilience policy combining insights from both insiders and outsiders. *Cities (London, England)* 90, pp. 157–167.

Lebel, L., Manuta, J.B. and Garden, P. 2011. Institutional traps and vulnerability to changes in climate and flood regimes in Thailand. *Regional Environmental Change* 11(1), pp. 45–58.

Lim, H. S., & Lu, X. X. 2016. Sustainable urban stormwater management in the tropics: An evaluation of Singapore's ABC Waters Program. *Journal of Hydrology, 538*, 842-862.

Loo, Y. Y., Billa, L., & Singh, A. 2015. Effect of climate change on seasonal monsoon in Asia and its impact on the variability of monsoon rainfall in Southeast Asia. *Geoscience Frontiers, 6*(6), 817-823.

Lotteau, M., Loubet, P., Pousse, M., Dufrasnes, E., & Sonnemann, G. 2015. Critical review of life cycle assessment (LCA) for the built environment at the neighborhood scale. *Building and Environment, 93*, 165-178.

Mancini, F. 2016. Commentary: Asian Cities: Fragility and Resilience. *Browser Download This Paper.*

Manuta, J., Khrutmuang, S., Huaisai, D., & Lebel, L. 2006. Institutionalized incapacities and practice in flood disaster management in Thailand.

Marks, D. 2015. The urban political ecology of the 2011 floods in bangkok: the creation of uneven vulnerabilities. *Pacific affairs* 88(3), pp. 623–651.

Marks, D. and Lebel, L., 2016. Disaster governance and the scalar politics of incomplete decentralization: Fragmented and contested responses to the 2011 floods in Central Thailand. Habitat International, 52, pp.57-66.

Marome, W. A. 2016. Enhancing Adaptation to Climate Change by Impact Assessment of the Flood in Bangkok. *Journal of Architectural/Planning Research and Studies, 13*(2), 31-40.

Mataki, M., Koshy, K. and Nair, V. 2007. Top-down, bottom-up: Mainstreaming adaptation in Pacific island townships. *Climate change and adaptation.*

Masoud, F., Margolis, L., & Khirfan, L. 2014. Integrated Water-Management: Landscape Infrastructure & Urban Morphology in the Jordan River Watershed. *Accessed October, 25.*

McNeel, R. 2010. Grasshopper-Generative Modeling with Rhino, McNeel North America, Seattle, USA.

Mega, V. 1996. Our city, our future: towards sustainable development in European cities. *Environment and Urbanization, 8*(1), 133-154.

Metcalfe, P., Beven, K., Hankin, B. and Lamb, R. 2017. A modelling framework for evaluation of the hydrological impacts of nature-based approaches to flood risk management, with application to in-channel interventions across a 29-km² scale catchment in the United Kingdom. *Hydrological processes* 31(9), pp. 1734–1748.

Meyer, V. 2018. Decision support on flood management in complex urban settings. is risk assessment the right approach or do we need decision heuristics? In: Kabisch, S., Koch, F., Gawel, E., Haase, A., Knapp, S., Krellenberg, K., Nivala, J., and Zehnsdorf, A. eds. *Urban Transformations*. Future City. Cham: Springer International Publishing, pp. 363–373.

Miller, C. 2015. Fact Sheet. *Imprint* 62(5), p. 17.

Moloney, S. and Fünfgeld, H., 2015. Emergent processes of adaptive capacity building: Local government climate change alliances and networks in Melbourne. Urban Climate, 14, pp.30-40.

Montgomery, M.R. 2008. The urban transformation of the developing world. *Science* 319(5864), pp. 761–764.

Murakami, A., Zain, A.M., Takeuchi, K., Tsunekawa, A. and Yokota, S., 2005. Trends in urbanization and patterns of land use in the Asian mega cities Jakarta, Bangkok, and Metro Manila. Landscape and Urban Planning, 70(3-4), pp.251-259.

Nakayama, H., Shimaoka, T., Omine, K., Patsaraporn, P. and Siriratpiriya, O., 2013. Solidwaste management in Bangkok at 2011 Thailand floods. Journal of Disaster Research, 8(3), pp.456-464.

Nasongkhla, S., & Sintusingha, S. 2011. Global aerotropolis versus local aqua-community: conflicting landscapes in the extended Bangkok Metropolitan Region, Thailand. *The Sustainable World, 142,* 205.

Naumann, S., Anzaldua, G., Berry, P., Burch, S., Davis, M., Frelih-Larsen, A., & Sanders, M. 2011. Assessment of the potential of ecosystem-based approaches to climate change adaptation and mitigation in Europe. *Final report to the European Commission, DG Environment.*

Nilubon, P., Veerbeek, W. and Zevenbergen, C. 2016. Amphibious architecture and design: A catalyst of Opportunistic Adaptation? – case study bangkok. *Procedia - Social and Behavioral Sciences* 216, pp. 470–480.

Nilubon, P., Veerbeek, W. and Zevenbergen, C., 2019. Integrating Climate Adaptation into Asset Management Planning: Assessing the Adaptation Potential and Opportunities of an Urban Area in Bangkok. *International Journal of Water Resources Engineering,* 4(2), pp.50-65.

Nilubon, P., Veerbeek, W. and Zevenbergen, C., 2019. Decision Tree Method (Flexibility in Adaptation–FIA) for Evaluating the Flexibility of Flood Risk Adaptation Options in Lad Krabang, Thailand. *International Journal of Water Resources Engineering,* 5(1), pp.32-48.

Nilubon, P., 2019. Challenges in implementing an Opportunistic Adaptation approach in Bangkok. *International Journal of Water Resources Engineering,* 5(2), pp.28-45.

Ogle, M., Ndhlovu, S. and Aalbaek, V. 2013. Zambia floods MDRZM008: DREF review. *Geneva:* IFRC.

Ontl, T.A., Swanston, C., Brandt, L.A., Butler, P.R., D'Amato, A.W., Handler, S.D., Janowiak, M.K. and Shannon, P.D. 2017. Adaptation pathways: ecoregion and land ownership influences on climate adaptation decision-making in forest management. *Climatic Change,* pp. 1–14.

Ouyyanont, P. 2008. The Crown Property Bureau in Thailand and the crisis of 1997. *Journal of Contemporary Asia,* 38(1), 166-189.

Park, K., & Lee, M. H. 2019. The Development and Application of the Urban Flood Risk Assessment Model for Reflecting upon Urban Planning Elements. *Water,* 11(5), 920.

Pathirana, A., Radhakrishnan, M., Ashley, R., Quan, N. H., & Zevenbergen, C. 2018. Managing urban water systems with significant adaptation deficits—unified framework for secondary cities: part II—the practice. *Climatic Change,* 149(1), 57-74.

Patil, D.D., Wadhai, V.M. and Gokhale, J.A., 2010. Evaluation of decision tree pruning algorithms for complexity and classification accuracy. International Journal of Computer Applications, 11(2), pp.23-30.

Peuportier, B. L. P. 2001. Life cycle assessment applied to the comparative evaluation of single family houses in the French context. *Energy and buildings,* 33(5), 443-450.

Peuportier, B. 2008. Life Cycle Assessment applications in the building sector. *International Journal of Environmental Technology and Management,* 9(4), 334-347.

Phien-Wej, N., Giao, P.H. and Nutalaya, P., 2006. Land subsidence in bangkok, Thailand. Engineering geology, 82(4), pp.187-201.

Pimonsathean, Y. 2012. The crown property bureau and heritage conservation. *The Journal of the Siam Society, 100,* 103-122.

Poapongsakorn, N., & Meethom, P. 2013. Impact of the 2011 floods, and flood management in Thailand. *ERIA Discussion Paper Series, 34*, 2013.

Pryor, S. C., Scavia, D., Downer, C., Gaden, M., Iverson, L., Nordstrom, R., ... & Robertson, G. P. 2014. Midwest. Climate change impacts in the United States: The third national climate assessment. *In: Melillo, JM; Richmond, TC; Yohe, GW, eds. National Climate Assessment Report. Washington, DC: US Global Change Research Program: 418-440.*, 418-440.

Quinlan, J.R., 1987. Simplifying decision trees. International journal of man-machine studies, 27(3), pp.221-234.

Radhakrishnan, M. and Zevenbergen, C. 2014. Applied Flood Resiliency: A Method for Determining the Recovery Capacity for Fast Growing Mega Cities. *Conference on Urban*

Radhakrishnan, M., Nguyen, H.Q., Gersonius, B., Pathirana, A., Vinh, K.Q., Ashley, R.M. and Zevenbergen, C. 2017. Coping capacities for improving adaptation pathways for flood protection in Can Tho, Vietnam. *Climatic Change* 149(1), pp. 1–13.

Radhakrishnan, M., Pathirana, A., Ashley, R. and Zevenbergen, C. 2017. Structuring Climate Adaptation through Multiple Perspectives: Framework and Case Study on Flood Risk Management. *Water* 9(2), p. 129.

Radhakrishnan, M., Pathirana, A., Ashley, R.M., Gersonius, B. and Zevenbergen, C. 2018. Flexible adaptation planning for water sensitive cities. *Cities (London, England)* 78, pp. 87–95.

Raman, J. K., Alves, C. M., & Gnansounou, E. 2017. A review on moringa tree and vetiver grass–potential biorefinery feedstocks. *Bioresource technology*, 249.

Rattanakanlaya, K., Sukonthasarn, A., Wangsrikhun, S., & Chanprasit, C. 2016. A survey of flood disaster preparedness among hospitals in the central region of Thailand. *Australasian emergency nursing journal*, *19*(4), 191-197.

Ren, Z., Chen, Z. and Wang, X. 2011. Climate change adaptation pathways for Australian residential buildings. *Building and Environment* 46(11), pp. 2398–2412.

Revi, A. 2008. Climate change risk: an adaptation and mitigation agenda for Indian cities. *Environment and urbanization* 20(1), pp. 207–229.

Rijke, J., Farrelly, M., Brown, R., & Zevenbergen, C. 2013. Configuring transformative governance to enhance resilient urban water systems. *Environmental Science & Policy*, 25, 62-72.

Ronald, R. 2008. Between investment, asset and use consumption: The meanings of homeownership in Japan. *Housing Studies*, *23*(2), 233-251.

Rosen, K. H., & Krithivasan, K. (2012). Trees in: Discrete mathematics and its applications: with combinatorics and graph theory. Tata McGraw-Hill Education. pp-528-552

Rossi, B., Marique, A.-F., Glaumann, M. and Reiter, S. 2012. Life-cycle assessment of residential buildings in three different European locations, basic tool. *Building and Environment* 51, pp. 395–401.

Rozos, E., Makropoulos, C., & Maksimović, Č. 2013. Rethinking urban areas: an example of an integrated blue-green approach. *Water Science and Technology: Water Supply*, *13*(6), 1534-1542.

Salami, A. W., Mohammed, A. A., Abdulmalik, Z. H., & Olanlokun, O. K. 2014. Trend analysis of hydro-meteorological variables using the Mann-Kendall trend test: Application to the Niger River and the Benue sub-basins in Nigeria. *International Journal of Technology*, *5*(2), 100-110.

Sayers, P.B., Galloway, G.E. and Hall, J.W. 2012. Robust decision-making under uncertainty – towards adaptive and resilient flood risk management infrastructure. In: Sayers, P. B. ed. *Flood risk: planning, design and management of flood defence infrastructure*. ICE Publishing, pp. 281–302.

Sharma, D., & Tomar, S. 2010. Mainstreaming climate change adaptation in Indian cities. *Environment and Urbanization*, *22*(2), 451-465.

Shen, J. 2007. Scale, state and the city: Urban transformation in post-reform China. *Habitat international* 31(3–4), pp. 303–316.

Shyam, S. S., Shridhar, N., & Fernandez, R. 2017. Climate change and need for proactive policy initiatives in Indian marine fisheries sector. *Climate Change*, *3*(9), 20-37.

Storch, H., & Downes, N. K. 2011. A scenario-based approach to assess Ho Chi Minh City's urban development strategies against the impact of climate change. *Cities*, *28*(6), 517-526.

Tanner, T., Mitchell, T., Polack, E., & Guenther, B. (2009). Urban governance for adaptation: assessing climate change resilience in Ten Asian Cities. *IDS Working Papers*, *2009*(315), 01-47.

Singkran, N. and Kandasamy, J. 2016. Developing a strategic flood risk management framework for Bangkok, Thailand. *Natural Hazards* 84(2), pp. 933–957.

Sintusingha, S. 2006. Sustainability and urban sprawl: Alternative scenarios for a Bangkok superblock. *URBAN DESIGN International* 11(3–4), pp. 151–172.

Spaans, M., & Waterhout, B. 2017. Building up resilience in cities worldwide–Rotterdam as participant in the 100 Resilient Cities Programme. *Cities*, *61*, 109-116.

Stumpe, J. and Tielrooij, F., 2000. Waterbeleid voor de 21e eeuw: Geef water de ruimte en de aandacht die het verdient. Advies van de Commissie Waterbeheer 21e eeuw. Advies aan de Staatssecretaris van Verkeer en Waterstaat en aan de voorzitter van de Unie van Waterschappen.

Tedeschi, A. 2011. *Parametric architecture with Grasshopper®: primer*. Le Penseur.

Thanvisitthpon, N. and Shrestha, S. 2018. Urban Flooding and Climate Change: A Case Study of Bangkok, Thailand. ... *and Urbanization ASIA*.

Terdpaopong, K., Rickards, R.C. and Manapreechadeelert, P. 2018. The 2011 floods' impact on the Thai industrial estates' financial stability: a ratio analysis with policy recommendations. *Environment, Development and Sustainability*, pp. 1–24.

Thomalla, F., Lebel, L., Boyland, M., Marks, D., Kimkong, H., Tan, S. B., & Nugroho, A. 2018. Long-term recovery narratives following major disasters in Southeast Asia. *Regional Environmental Change*, *18*(4), 1211-1222.

Tingsanchali, T. 2012. Urban flood disaster management. Procedia engineering, 32, 25-37.

Toker, Z., 2007. Recent trends in community design: the eminence of participation. *Design Studies*, 28(3), pp.309-323.

Townend, I., Sayers, P.B., Hinton, C., Panzeri, M.C., Cooper, N., Nicholls, R.J., Thorne, C.R. and Simm, J., 2005. OST Foresight Report: A futures analysis of UK coastal flooding and erosion. *In International Conference on Coastlines, structures and breakwaters 2005*, pp. 1-18.

Van der Pol, J. 2011. Flood proof architecture. Climate of Coastal Cooperation. Netherlands: Coastal.

Vanno, Sirintra. "Bangkok's green infrastructure." Journal of Architectural/Planning Research and Studies (JARS) 9, no. 2 (2012): 1-14.

Van Schaick, J., & Klaasen, I. (2011). The Dutch layers approach to spatial planning and design: a fruitful planning tool or a temporary phenomenon?. European Planning Studies, 19(10), 1775-1796

Veerbeek, W., & Zevenbergen, C. 2009. Deconstructing urban flood damages: increasing the expressiveness of flood damage models combining a high level of detail with a broad attribute set. *Journal of Flood Risk Management*, 2(1), 45-57.

Veerbeek, W., Denekew, H., Pathirana, A., Brdjanovic, D., Zevenbergen, C., & Bacchin, T. K. 2011. Urban growth modeling to predict the changes in the urban microclimate and urban water cycle.

Veerbeek, W. and Ashley, R.M. 2012. Building adaptive capacity for flood proofing in urban areas through synergistic interventions. *WSUD 2012: Water*

Vercruysse, K., Dawson, D.A. and Wright, N., 2019. Interoperability: A conceptual framework to bridge the gap between multifunctional and multisystem urban flood management. *Journal of Flood Risk Management*, 12(S2), p.e12535.

Wagdy, A., Elghazi, Y., Abdalwahab, S., & Hassan, A. 2015. The balance between daylighting and thermal performance based on exploiting the kaleidocycle typology in hot arid climate of Aswan, Egypt. In *AEI 2015* (pp. 300-315).

Walker, W.E., Rahman, S.A. and Cave, J., 2001. Adaptive policies, policy analysis, and policy-making. European journal of operational Research, 128(2), pp.282-289.

Wei, H., Chengyi, T. and Renjie, Z. 2019. Research on flood discharge, energy dissipation, and operation mode of sluice gate for low-head and large-discharge hydropower stations. ... *and Safe Dams Around the World*

Werner, M.G.F., 2001. Impact of grid size in GIS based flood extent mapping using a 1D flow model. Physics and Chemistry of the Earth, Part B: Hydrology, Oceans and Atmosphere, 26(7-8), pp.517-522.

Wise, R.M., Fazey, I., Stafford Smith, M., Park, S.E., Eakin, H.C., Archer Van Garderen, E.R.M. and Campbell, B. 2014. Reconceptualising adaptation to climate change as part of pathways of change and response. *Global Environmental Change* 28, pp. 325-336.

Wong, T. H. F., & Brown, R. R. 2009. The water sensitive city: principles for practice. *Water Science & Technology*, 60(3).

Woodward, M., Kapelan, Z. and Gouldby, B. 2014. Adaptive flood risk management under climate change uncertainty using real options and optimization. *Risk Analysis* 34(1), pp. 75–92.

World Bank Group. 2017. Sovereign climate and disaster risk pooling.

Wynes, S., & Nicholas, K. A. 2017. The climate mitigation gap: education and government recommendations miss the most effective individual actions. *Environmental Research Letters, 12*(7), 074024.

Yin, H. and Li, C. 2001. Human impact on floods and flood disasters on the Yangtze River. *Geomorphology* 41(2–3), pp. 105–109.

Yokohari, M., Takeuchi, K., Watanabe, T., & Yokota, S. 2000. Beyond greenbelts and zoning: A new planning concept for the environment of Asian mega-cities. *Landscape and urban planning, 47*(3-4), 159-171.

Zandvoort, M., Campos, I.S., Vizinho, A., Penha-Lopes, G., Lorencová, E.K., van der Brugge, R., van der Vlist, M.J., van den Brink, A. and Jeuken, A.B.M. 2017. Adaptation pathways in planning for uncertain climate change: Applications in Portugal, the Czech Republic and the Netherlands. *Environmental Science & Policy* 78, pp. 18–26.

Zevenbergen, C., Veerbeek, W., Gersonius, B., Thepen, J., & Van Herk, S. 2008. Adapting to climate change: using urban renewal in managing long-term flood risk. *WIT Transactions on Ecology and the Environment, 118*, 221-233.

Zevenbergen, C., Veerbeek, W., Gersonius, B., & Van Herk, S. 2008. Challenges in urban flood management: travelling across spatial and temporal scales. *Journal of Flood Risk Management, 1*(2), 81-88.

Zevenbergen, C., Cashman, A., Evelpidou, N., Pasche, E., Garvin, S., & Ashley, R. 2010. *Urban flood management.* CRC Press.

Zevenbergen, C., Rijke, J. S., Van Herk, S., Ludy, J., & Ashley, R. 2013. Room for the River: International relevance. *Water Governance, 3 (2).*

Zevenbergen, C., van Herk, S., Rijke, J., et al. 2013. Taming global flood disasters. Lessons learned from Dutch experience. *Natural Hazards* 65(3), pp. 1217–1225.

Zevenbergen, C., van Herk, S., Escarameia, M., Gersonius, B., Serre, D., Walliman, N., & de Graaf, R. 2018. Assessing quick wins to protect critical urban infrastructure from floods: a case study in Bangkok, Thailand. *Journal of Flood Risk Management, 11*, S17-S27.

Zheng, H.W., Shen, G.Q. and Wang, H. 2014. A review of recent studies on sustainable urban renewal. *Habitat international* 41, pp. 272–279.

Appendix A:
Details case studies dynamic design (Chapter 4)

Workshops and competitions 2018 - 2020: Lad Kraband district, Thailand. Three cases for dynamic design are presented, which different scales such as autonomous adaptation and Top-down adaptation.

Case studies 1 and 2 were presented to the workshop and local community by local authorities. Case study 3 is a competition project designed by the local architecture company (nppn-company), presenting novel flood adaptation measures.

Case study 1: Autonomous adaptation

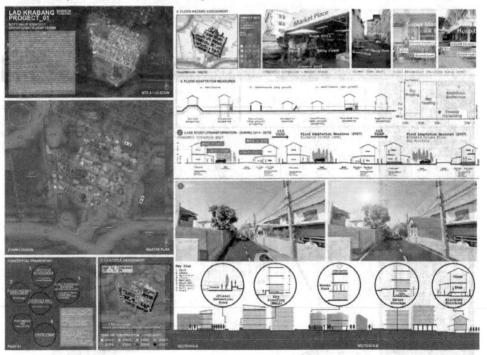

Case study 2: Top-down adaptation (Design approach)

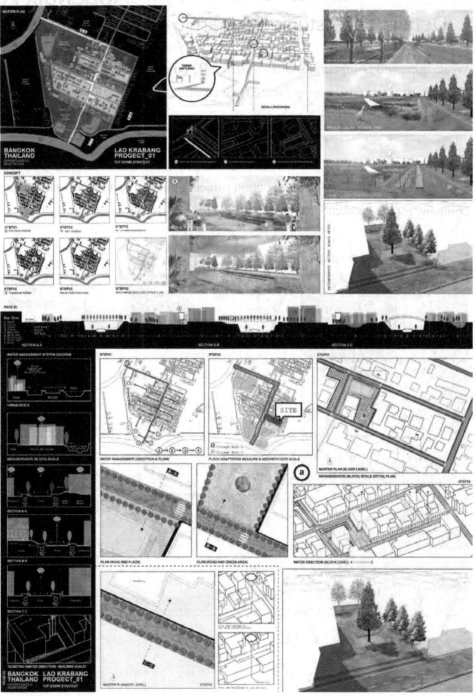

Current situation at Kehanakorn village 2

First implementation – flood walls

Second implementation – elevated road system

Last implementation - floodways

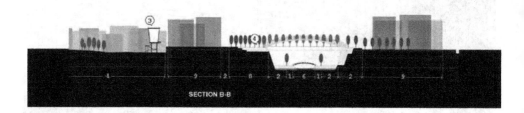

SECTION B-B

Case study 3 - Amphi-Garden (Amphibious + Garden)

We!park Competition 2020: Ekkamai Pocket - Bangkok, Thailand

Location: Ekkamai, Bangkok

The design approach (Amphi-Graden) proposes a platform to transform and connect 'empty' spaces to become a "green-blue commonality space"

Concept

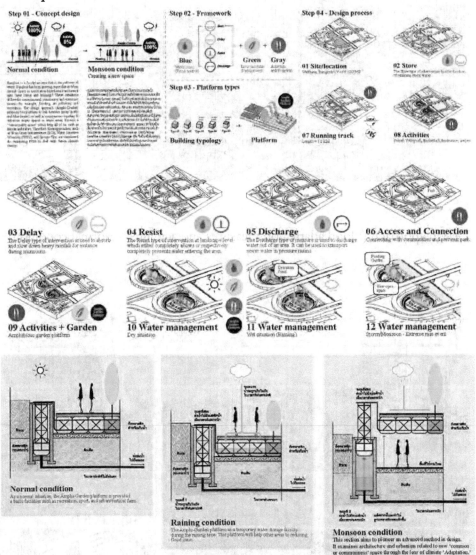

Current situation

Design approach (Flood adaptation measures and activities – normal condition)

Design approach (monsoon condition) – floating platform

Details: Long-section

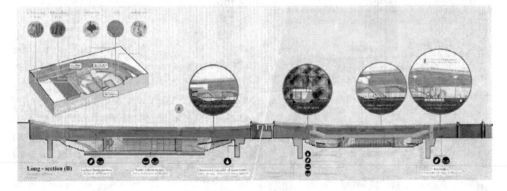

Perspective 1

Perspective 2

Perspective 3 - New space under (amphi garden platform).

Master plan

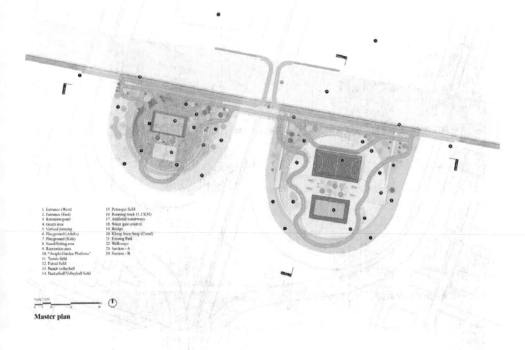

1. Entrance (West)
2. Entrance (East)
3. Retention pond
4. Green area
5. Vertical farming
6. Playground (Adults)
7. Playground (Kids)
8. Stand/Sitting area
9. Recreation area
10. "Amphi-Garden Platform"
11. Tennis field
12. Futsal field
13. Beach volleyball
14. Basketball/Volleyball field
15. Petanque field
16. Running track (1.2 KM)
17. Artificial waterways
18. Water gate control
19. Bridge
20. Khlong Saen Saep (Canal)
21. Existing Park
22. Walkways
23. Section - A
24. Section - B

Master plan

Appendix B:
Details Pathways – Dynamic design (one cluster)

This section presents a dynamic design based on three pathways as below: Pathways 1 (landscape level), Pathways 2 (architecture and landscape levels), and Pathways 3 (architecture level).

1. Pathways 1

These figures visualize pathway 1. The adaptation measure to be implemented in 2032 is based on simulation. Pathway 1 is a single opportunity to adapt, namely, in 2032. Since the measure is applied at the landscape level, it has the potential to transform the space during the dry season allowing for leisure activities, fitness, etc. Whereas, during the rainy season, the effect of the measure is to use the area as a retention pond, able to store water during local flood events.

An example illustration of one cluster in Pathways 1

2. Pathways 2

Pathway 2 is visualized in figure 1 to 4. Figure (1) presents the current situation. Figure (2) (first adaptation) shows the "Water square" in the year 2032, which allows water to be stored in the area. This measure provides multiple functions in the area which are dependent of the season: dry or rainy. A playground, a plaza, or a football field could be there in the dry season, while other activities such as fishing could be planned for the rainy season. The lifespan of this measure ends approximately after 13 years. Figure (3) presents the second measure, which is called "Adding Green in the Streetscape" and the year of implementation is 2044. This measure is part of the Delay component and is using the green space to absorb and delay rainwater before it is filling the water square. The measure also allows to integrate additional functions such as a walkway and a bike lane. This measure could be combined with the previous one (water square) to allow the citizens to share a common space with the water. Finally, a third measure involves elevating the (floor levels of the) building. Figure (4) presents all the three measures combined. The water square becomes a water plaza for children to enjoy. During the dry season, the water square is used to store water to use for watering the green areas. These three measures together enhance the livability and urban quality of the area.

An example illustration of one cluster in Pathways 2

3. Pathways 3

Pathway 3 visualizes pathway 3 and presents measures to be all implemented at the architecture level, involving a tank roof, a garden, or an elevated ground level of the building. Figure (1) presents the current situation. Figure (2) to (4) visualizes the first, second and third measure to be implemented over time in 2032, 2040 and 2047, respectively. (Raising the floor level (only the building) presents the last intervention around 2047, which is 7 years later than the second intervention.

Current (2020) ⟶ Tank roof (2032) ⟶ Gardens (2040) ⟶ Raising the ground flood level (2047) ⟶ End (2062)

An example illustration of one cluster in Pathways 3

Example perspective (one cluster)

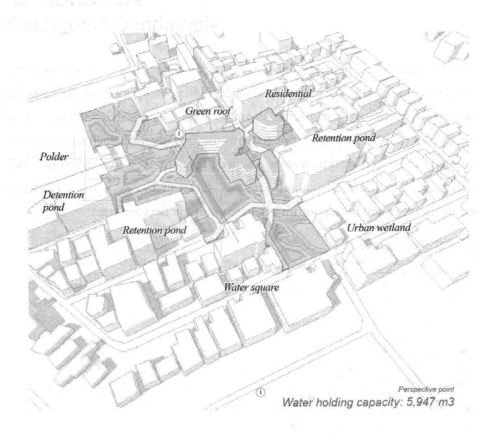

Residential

Green roof

Polder

Retention pond

Detention
pond

Retention pond

Urban wetland

Water square

Perspective point
Water holding capacity: 5,947 m3

Appendix C:
Algorithm OA approach

The algorithm that generates a decision tree is a recursive one. This means that it operates in the same way at each individual decision point by applying all adaptation measures that fit to that specific situation. This creates an additional set of possibilities adaptation points connected to the first one on which the process is repeated until a certain time limit is reached. This creates a tree structure which represents all possible adaptation pathways that can be taken from the initial starting point where the algorithm began (root of the decision tree).

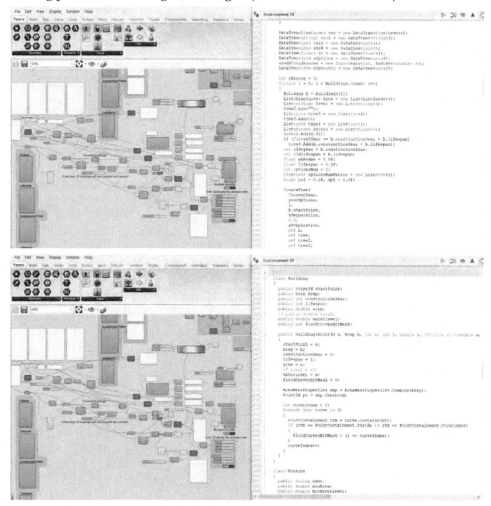

These images present the algorithm that generates a decision tree.

QR code (1)

Scan me – This shows the result of the LCA: using the urban renewal cycle for climate change which based on algorithm. (Autonomous adaptation) - Lad Krabang district.

Appendix D:
Catalog of flood adaptation measures

Flood Adaptation Measures	Area Requirements (Sq. Meters)	Lifespan (Year)	Water Level (Meters)	Flood Frequency (Year)
Resist (Architecture)				
Sealable buildings (dry proof)	10	12	0.3	Every 1
Wet proofing (water resistant construction)	10	15	0.6	Every 1
Protection life support facilities and dangerous goods	250	50	2	Every 50
Use of buildings as flood defence	150	50	1.5	Every 2
Resist (Landscape)				
compartments in dike rings	1,000	20	0.3	Every 20
Elevated flood wall	500	10	0.6	Every 1
Dikes	2,000	15	2	Every 1
Floodable dike	2,000	50	1.5	Every 20
Floodplain excavation of enlargement	4,000	20	0.6	Every 1
Overtopping-proof dike	1,500	50	1	Every 1
Super dike	4,500	50	2	Every 2
Unbreakable dike	3,000	50	2	Every 2
Avoid (Architecture)				
Elevated building	50	20	3	Every 1
Increase height difference between street and ground	30	20	1	Every 5
Raising the ground floor level	20	15	0.6	Every 1
Buildings (partly) Situated in the water	100	30	0.6	Every 1
Floating buildings	50	50	2	Every 5
Relocation of buildings, utility facilities and infrastructure	250	50	3	Every 1
Safe ground for flood events	200	30	1	Every 2
Amphibious (floatable) constructions	50	50	1.5	Every 1
Construction on piles	50	25	3	Every 10
Avoid (Landscape)				
Building on partially elevated areas	2,000	25	2	Every 1
Raising land	1,000	5	3	Every 1
Artificial islands	4,000	30	3	Every 2
Evacuation routes at elevated level	1,000	20	1.5	Every 1
Dismountable and temporary buildings	500	20	0.6	Every 1
Raised curbs/hollow roads	1,000	50	0.3	Every 2
Delay (Architecture)				
Sloping roof	25	10	0.3	Every 1
Green facades	10	5	0.3	Every 1
Green roof	10	5	0.6	Every 1
Wetting surface (gardens, roads)	30	10	0.6	Every 2

Delay (Architecture)			
Sloping roof	25	10	0.3	Every 1
Green facades	10	5	0.3	Every 1
Green roof	10	5	0.6	Every 1
Wetting surface (gardens, roads)	30	10	0.6	Every 2
Delay (Landscape)				
Pocket park	850	10	0.3	Every 1
Green infrastructure	1,000	20	0.5	Every 2
Urban wetland	4,000	20	0.6	Every 1
linear park	850	10	0.3	Every 1
Park	1,000	12	0.3	Every 10
Green shores and riverbanks	10,000	40	2	Every 1
infiltration fields	2,000	8	2	Every 1
improve soil infiltration capacity	1,000	10	2	Every 1
Wadi	2,500	30	2	Every 2
shallow infiltration measures	500	10	1.5	Every 10
Adding green in streetscape	500	7	0.6	Every 1
Inclination of roads	2,000	15	1	Every 2
Drainage below surface level	2,000	5	0.6	Every 2
Porous pavements	100	10	0.6	Every 1
Store (Architecture)				
Airbag water storage	10	15	0.3	Every 1
Rainwater tanks	5	10	0.3	Every 1
Tank roof	20	10	0.6	Every 1
Store (Landscape)				
Water Square (plaza)	500	30	2	Every 1
By-pass creation	3,500	30	1	Every 1
retention area	4,000	20	1.5	Every 1
polder	2,000	15	0.6	Every 1
increased storage of discharge capacity of surface water	750	20	0.3	Every 1
Deepen water bodies	1,500	25	0.3	Every 2
Ditches	2,000	20	0.5	Every 1
creating swimming location	500	20	0.3	Every 10
Water basins	5,500	25	0.3	Every 20
Increased storage	1,500	30	0.3	Every 1
Dams	5,000	50	2	Every 20

www.climateapp.org/

Appendix E:
Survey perception flood risk of Bangkok

General flood management issue (Interview)	Pre-Question (Questionnaire)	Post-Question (Questionnaire)
1. Did this disaster (flooding 2011) cause a change in our attitude to climate adaptation or not? Beyond technical and financial challenges, one of the main issues in climate change adaptation is how the process is facilitated. **2.** How is climate adaptation currently organized and how is its urgency perceived by different stakeholders.	**1.** Currently, people in urban flood risk areas have gathered better knowledge and understanding about the flood situation, as well as how it should be dealt with from the perspective of the government.	**1.** This new approach can provide knowledge and understanding about problems and methods to solve flood problems to people in the area as well.
Knowledge & awareness **1.** Do you feel that the flood risk is essential and urgent? **2.** Do you know how to deal with the flood situation? **3.** What options are the governors providing to the local people? **4.** How to provide knowledge and awareness to the local people?	**2.** At present, the government has already shown to the public, which are the best available solutions that can effectively solve the flood issue in the affected areas.	**2.** This new approach can be supervised. It is important to have effective flood solutions which can be supervised and thus also accepted by all parties, such as government officials and local people
Mode of Operation **1.** How should the operation proceed in the given area? **2.** Is there an established strategy on adaptation to climate change being implemented in Bangkok? **3.** What is the problem of the previous method? **4.** How has the current method operated?	**3.** The organization and the operation processes are straightforward and uncomplicated.	**3.** This new approach provides a simple and uncomplicated procedure for solving flood problems in the area.
Cooperation & Governance **1.** How does the local government cooperate with the local community? **2.** How have the people organized themselves to deal with the flood situation?	**4.** Regulations and policies intended as flood solutions are being effectively used by the government. For example, the building regulation does not allow the construction to encroach into the public water storage or discharge areas such as canals.	**4.** This new approach is more likely to request funding to solve flood problems in the area more easily, by requesting financial support from the public and private sectors

Economic & Financial Requirements

1. How does the community find the financial support offered for solving flood situation?

2. How to secure the funding for each project?

Additional Benefits

1. What are the multiple benefits that are caused by applying flood adaptation measures?

2. Is it possible for the current method to offer multiple benefits to the local people and to the area of application?

3. What are the benefits that flood adaptation measures can provide as a side-effect of the initial flood solution offered by the government?

5. Nowadays, finding and securing funding for solving flood related issues, such as out of the budget of a private company or out of the national budget is very easy.

6. Flood measures, such as the giant's tunnel, provide many benefits for the physical areas/regions. For instance, there might be an increase in the livability of the area and in the general health of the inhabitants.

7. Most of the flood solutions that are decided by the government can be better and can more efficiently deal with current flood events.

8. Current strategies can probably be dealing with future flood events.

5. This approach of solving the flood problem using the new platform is very beneficial to the people in the area.

6. This new approach can provide various guidelines in solving flood problems for people in the area as well.

7. Guidelines on how to solve the flood problem in this new format can be used effectively, especially in areas affected by flooding.

8. This new approach can cope with flooding that will occur in the future as well.

About the author

Polpat Nil-u-bon was born on 13 May 1985 in Bangkok, Thailand. He has a university degree in Architecture (2008 – B.Arch.) from Rajamangala University of Technology (RMUTT), Thailand and two Master degrees in Architectural and Environmental Technology (2010) from Chulalongkorn University (CU), Thailand and in Architecture and Urban Design from The Berlage Institute and TU Delft, the Netherlands (2013).

After graduation in 2008, he worked for three years at HASSELL in Thailand, before entering The Berlage Institute. In 2011 he got a scholarship for a Master degree and Ph.D. study in Architecture and Urbanism at The Berlage Institute and IHE Delft by Rajamangala University of Technology (RMUTT), Thailand.

Currently, he is a full-time Ph.D. student research fellow of Flood Resilience Group (FRG – started 2014), Water Science & Engineering (WSE) at IHE Delft: Institute for Water Education, the Netherlands. The title of his research is "Opportunistic Adaptation: using the urban renewal cycle for climate change". The research project is funded by Rajamangala University of Technology (RMUTT).

Education

2014 – Present: Ph.D. Research: Opportunistic Adaptation: using the urban renewal cycle to adapt to climate change, IHE Delft: Institute for Water Education, Delft, the Netherlands, and TUDelft: Delft University of Technology, Delft, the Netherlands.

2011 - 2013:   Master of Excellence in Architecture, Urban planning and Landscape Design - 2nd The Berlage: Center for Advanced Studies in Architecture and Urban Design, (TUDelft, Delft, the Netherlands), Dissertation: Siamese Metropolises - A cartography of relations between Shenzhen and Hong-Kong; Fish City.

2008 - 2010:   Master of Architecture (Architectural and Environmental Technology) - 1st Chulalongkorn University (CU), Bangkok, Thailand, Dissertation: The Development of Building Materials from Agricultural Residues.

2003 - 2008:   Bachelor of Architecture (Architecture Technology): Rajamangala University of Technology, Pathumthani, Thailand - Dissertation: Vertical Park - Green building for reducing pollution

2008:   Memberships: Member of the Association of Siamese Architects (ASA)

Experience record

1. Moderator, workshop (UN-HABITAT and BMA), SDG project assessment tailoring workshop, 18 February 2020.
2. Organization and presentation, workshop (Water Management for Urban Development Planning in the Bangkok Metropolitan Region, 7-18 November 2016)
3. A committee of the conference, organization, and operation - ICAADE 2015 is the very first International Conference on Amphibious Architecture, Design, and Engineering Bangkok, Thailand, 26-29 August 2015.

4. Lecturer: Faculty of Architecture, Rajamangala University of Technology, Design Studio second year, 2010.

5. Nominee and Participation, Thailand. Archiprix International Montevideo Workshops 2009, Uruguay: World's best graduation projects 2009 (Architecture International Competition) "Vertical Park" During March 26 - April 10, 2009

6. Has successfully completed the higher education course on "lighting system design and simulation" During February 6-28, 2009

7. Assistant and Coordinator's Dr.Acharawan Chutarat, Faculty of Architecture, King Mongkut's University of Technology Thonburi: Luxpacifica 6 "Light without Border" During April 23-25, 2009

8. Assistant's Dr.Vorapat Inkarojrit, Faculty of Architecture, Chulalongkorn University: "Evaluation of the Impact of Tuberculosis Infection Control Measures in Thai Hospital" (Evaluation of Environmental) During November 2009

Journal Papers

1. Nilubon, P., Veerbeek, W. and Zevenbergen, C. 2016. Amphibious architecture and design: A catalyst of Opportunistic Adaptation? – case study Bangkok. *Procedia - Social and Behavioral Sciences* 216, pp. 470–480.

2. Nilubon, P., Veerbeek, W. and Zevenbergen, C., 2019. Integrating Climate Adaptation into Asset Management Planning: Assessing the Adaptation Potential and Opportunities of an Urban Area in Bangkok. *International Journal of Water Resources Engineering, 4*(2), pp.50-65.

3. Nilubon, P., Veerbeek, W. and Zevenbergen, C., 2019. Decision Tree Method (Flexibility in Adaptation–FIA) for Evaluating the Flexibility of Flood Risk Adaptation Options in Lad Krabang, Thailand. *International Journal of Water Resources Engineering, 5*(1), pp.32-48.

4. Nilubon, P., 2019. Challenges in implementing an Opportunistic Adaptation approach in Bangkok. *International Journal of Water Resources Engineering, 5*(2), pp.28-45.

5. Nilubon, P., 2013. Siamese Metropolises: Relational Urbanism Studio. *The Berlage 2013*, CRC Press.

6. Nilubon, P., 2012. Plans and Platitudes: City of blind. *Journal of Berlage Institute*, 2(1), pp. 35-40.

7. Nilubon, P., 2010. The Development of building materials from agricultural residues. *Journal of Chulalongkorn University*, 3(2), pp.20-23.

Conferences

1. Nilubon, P., 2018. Land and Water Management in Bangkok Metropolitan Region 2018.

2. Nilubon, P., 2016. Opportunistic Adaptation: using the urban renewal cycle for climate change. Water management for urban development planning in the Bangkok metropolitan region (BMR) 2016.

3. Nilubon, P., 2016. Student Workshop: Adapting cities to climate change using urban renewal. PhD symposium "Urban sustainability 2016".

4. Nilubon, P., 2015. Opportunistic Adaptation of Architecture and Urbanism. *ICAADE 2015*.

5. Nilubon, P., 2015. Amphibious Architecture and Design: A Catalyst of Opportunistic Adaptation? – Case Study Bangkok. Urban Planning and Architectural Design for Sustainable Development *(UPADSD) 2015*.

6. Nilubon, P., 2014. Opportunistic Adaptation of Architecture and Urbanism. *Urban adaptation to climate change*, Deltas in Times of Climate Change II - Rotterdam, The Netherlands 24 - 26 Sep.

7. Nilubon, P., 2009. Archiprix International, (2009). International: 010 PUBLISHERS.

*Netherlands Research School for the
Socio-Economic and Natural Sciences of the Environment*

DIPLOMA

for specialised PhD training

The Netherlands research school for the
Socio-Economic and Natural Sciences of the Environment
(SENSE) declares that

Polpat Nilubon

born on 13 May 1985 Bangkok, Thailand

has successfully fulfilled all requirements of the
educational PhD programme of SENSE.

Delft, 28 April 2021

The Chairman of the SENSE board

Prof. dr. Martin Wassen

the SENSE Director of Education

Dr. Ad van Dommelen

The SENSE Research School has been accredited by the Royal Netherlands Academy of Arts and Sciences (KNAW)

KONINKLIJKE NEDERLANDSE
AKADEMIE VAN WETENSCHAPPEN

The SENSE Research School declares that Polpat Nilubon has successfully fulfilled all
requirements of the educational PhD programme of SENSE with a
work load of 37.7 EC, including the following activities:

SENSE PhD Courses

o   Environmental research in context (2014)
o   SENSE writing week (2014)
o   Research in context activity: 'Co-organizing ICAADE 2015: First International Conference
    on Amphibious Architecture, Design and Engineering, 26-29 August 2015, Bangkok,
    Thailand'

Other PhD and Advanced MSc Courses

o   Urban flood management, TU Delft (2014)
o   Turning your thesis into business, TU Delft (2018)
o   Academic writing, IHE Delft (2018)

Management and Didactic Skills Training

o   Organising and managing the student workshop 'Adapting cities to climate change using
    urban renewal', 29 September, 2014 Delft, The Netherlands
o   Organising student design workshop 'Amphibious Architecture Design and Engineering
    Design Workshop', 20-25 August 2015 , in Bangkok, Thailand
o   Organising the workshop 'Water Management for Urban Development Planning in the
    Bangkok Metropolitan Region', 7-18 November 2016

Oral Presentations

o   *Opportunistic Adaptation of Architecture and Urbanism.* ICAADE 2015 - International
    conference on amphibious architecture, design & engineering, 26-29 August 2015,
    Bangkok, Thailand
o   *Amphibious Architecture and Design: A Catalyst of Opportunistic Adaptation? – Case
    Study Bangkok.* Urban Planning and Architectural Design for Sustainable Development,
    15 October 2015, Lecce, Italy
o   *Opportunistic Adaptation: using the urban renewal cycle for climate change.* Water
    management for urban development planning in the Bangkok metropolitan region
    (BMR), 29 September 2016, Bangkok, Thailand

SENSE coordinator PhD education

Dr. ir. Peter Vermeulen

Urban climate adaptation currently focusses mainly on hazards but often ignores opportunities which arise in both space and time. Opportunistic Adaptation provides a rationalized approach to mainstream measures for climate adaptation into urban renewal cycles. Adaptation opportunities are identified by projecting the lifespans of urban assets into the future to obtain an operational urban adaptation agenda for the future. Upscaling of the adaptation process is done by synchronizing the end-of-lifecycle of a group of assets to develop adaptation clusters that comprise multiple dwellings, infrastructure as well as public spaces. An extensive catalogue of adaptation measures for different scale-levels ensures flexibility in the type of measures that can be integrated.

Sequencing the adaptation measures over long periods of time provides insight and flexibility in the long-term protection standards that can be achieved. By applying a design-centered approach, the potentials of obtaining co-benefits in the urban landscape are maximized. Potentials of clustering of nature-based solutions are being considered which ensures to maximize the delivery of ecosystem services.

This research aims to assess \the adaptation potential of Bangkok, based on a case study area (Lat Krabang) by mapping the adaptation opportunities and flood vulnerability. The resulting outputs will contribute to the development of a flexible and inclusive FRM strategy.

A BALKEMA BOOK

This book is printed on paper from sustainably managed forests and controlled sources

9 781032 055091

an **informa** business